Lecture Notes of
the Unione Matematica Italiana

27

More information about this series at http://www.springer.com/series/7172

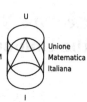

Enrico Carlini • Huy Tài Hà • Brian Harbourne •
Adam Van Tuyl

Ideals of Powers and Powers of Ideals

Intersecting Algebra, Geometry, and Combinatorics

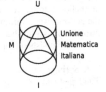

Unione
Matematica
Italiana

Enrico Carlini
Department of Mathematical Sciences
Politecnico di Torino
Torino, Italy

Huy Tài Hà
Department of Mathematics
Tulane University
New Orleans, LA, USA

Brian Harbourne
Department of Mathematics
University of Nebraska
Lincoln, NE, USA

Adam Van Tuyl
Department of Mathematics and Statistics
McMaster University
Hamilton, ON, Canada

ISSN 1862-9113   ISSN 1862-9121   (electronic)
Lecture Notes of the Unione Matematica Italiana
ISBN 978-3-030-45246-9   ISBN 978-3-030-45247-6   (eBook)
https://doi.org/10.1007/978-3-030-45247-6

Mathematics Subject Classification: 13-02, 14-02, 11P05, 14Q05, 13D40, 13F55, 13F20, 14C20

This Springer imprint is published by the registered company Springer Nature Switzerland AG.
The registered company address is: Gewerbestrasse 11, 6330 Cham, Switzerland

We dedicate this book to all of the participants who have made PRAGMATIC such a success for 20 years and to the Department of Mathematics at the University of Catania, for providing such a supportive research environment.

Foreword

The PRAGMATIC research school began in 1997 with its first edition. Since the last 20 years, many of the most important researchers in the field of Algebraic Geometry and Commutative Algebra have succeeded as main teachers of this project and have made this school a real success among young researchers (see the website of PRAGMATIC: www.dmi.unict.it/pragmatic/docs/Pragmatic-main.html).

In addition, the almost 400 young researchers who participated in the various editions of PRAGMATIC have in turn contributed to the success of this activity with over one hundred publications in the most important scientific journals (a partial list of such works can be found at www.dmi.unict.it/~pragmatic/docs/Pragmatic-papers.html). In thanking, on behalf of the organizers, all those who contributed to the enormous success of PRAGMATIC at the end of these few lines, the list of main speakers and collaborators who have been present over the years is included. To make this school of research even more popular, the 2017 edition has been enriched by the present nice volume that collects the lessons held during the period of the school and all the material necessary to address the open problems proposed in this edition. The authors, all of whom were involved in Pragmatic 2017, had the wonderful idea of putting together both the preparatory material, the lessons made during the school period, the open problems assigned and the state of the art at the end of the three weeks of activity.

Already the intriguing title announces in an original and curious way the field of research that is treated. Precisely, in this book powers of ideals and ideals of powers are approached from three different points of view: algebra, combinatorics, and geometry, with special regard to the interactions among these perspectives. I believe that these notes will be useful not only for the group of participants in the 2017 edition of PRAGMATIC but also for all those who have interests in the study of the powers of ideals and their applications in different fields of mathematics. Moreover, the way the text is composed also provides a simple and fruitful way to approach open issues in this area. The organizers of PRAGMATIC are very grateful to Enrico Carlini, Huy Tài Hà, Brian Harbourne, and Adam Van Tuyl for making this additional effort to be useful to the young mathematical community and for doing so in the context of PRAGMATIC.

Here is the list (in a chronological order) of the main teachers involved in all the editions of PRAGMATIC: David Eisenbud, Lawrence Ein, Anthony V. Geramita, Juan Migliore, Klaus Hulek, Kristian Ranestad, Ciro Ciliberto, Rick Miranda, Fyodor Zak, Massimiliano Mella, Igor Dolgachev, Alessandro Verra, Olivier Debarre, Lucia Caporaso, Lucian Badescu, Francesco Russo, Giuseppe Pareschi, Mihnea Popa, Jurgen Herzog, Volkmar Welker, Rosa Miró Roig, Giorgio Ottaviani, Paltin Ionescu, Jaroslav A. Wisnieswski, Ralf Fröberg, Mats Boij, Alessio Corti, Paolo Cascini, Yujiro Kawamata, Aldo Conca, Srikanth Iyngar, Anurag Singh, Alessandro Chiodo, Filippo Viviani, Gian Pietro Pirola, Joan Carles Naranjo, Brian Harbourne, and Adam Van Tuyl.

And here is a list of young collaborators involved during these years: S. Popescu, V. Masek, A. Bigatti, C. Peterson, F. Flamini, A. Bruno, G. Pacienza, C. Casagrande, A. Rapagnetta, M. Vladiou, X. Zheng, E. Nevo, L. Costa, D. Faenzi, L. Sola Condé, J.C. Sierra, V. Crispin, A. Engström, A. Kasprzyk, Y. Gongyo, G. Codogni, J. Guéré, L. Stoppino, V. Gonzalez-Alonso, E. Carlini, and Huy Tài Hà.

Catania, Italy   Alfio Ragusa
September 2018

Preface

This book contains reorganized and extended versions of our lectures at PRAG-MATIC 2017, held from June 19th to July 7th, 2017. PRAGMATIC (Promotion of Research in Algebraic Geometry for MAThematicians in Isolated Centres) is an annual summer school, started in 1997 and organized by the Università di Catania, Catania, Italy. The goal of the school is to stimulate research in algebraic geometry among Ph.D. students and early career researchers, especially those living in isolated centers or peripheral universities all over Europe.

We celebrate PRAGMATIC's twentieth anniversary with the theme "powers of ideals and ideals of powers." This theme became the title of our book. The topics in this book are motivated by algebraic problems involving the relationship between various notions of powers of ideals and by related geometric problems. This includes, as just one example among a constellation of variations, the Waring problem of writing a homogeneous polynomial as a minimal sum of powers of linear homogeneous polynomials, which translates algebraically into studying ideals of powers and geometrically into studying dimensions of secant varieties.

In these notes, powers of ideals and ideals of powers are approached from three points of view—algebra, combinatorics, and geometry—and the interactions between these perspectives will be developed. Readers are invited to explore the evolution of the set of associated primes of higher and higher powers of an ideal. For ideals associated with a combinatorial object like a graph or hypergraph, one wishes to explain this evolution in terms of the original combinatorial objects. Similar questions concern understanding the Castelnuovo–Mumford regularity of powers of combinatorially defined ideals in terms of the associated combinatorial data. From a more geometric point of view, one can consider how the relations between symbolic and regular powers can be interpreted in geometrical terms. Other topics to be presented include aspects of Waring type problems, symbolic powers of an ideal and their invariants (e.g., the Waldschmidt constant, the resurgence), and the persistence of associated primes.

When preparing our lectures for PRAGMATIC, our emphasis was on quickly introducing the participants to open problems and questions in these research areas. At the same time, we wanted to provide restricted versions of these problems and questions focused on specific cases which participants, with a minimal background in these areas, could tackle. Our intention was for these specific cases to be simple enough that participants would make significant progress within the 3-week school time frame, and yet important enough that their work could lead to publications and further investigation of these problems and questions in more generality. With this in mind, our focus was on the context of the problems and on how problems, results, and methods have evolved. Consequently, our lecture notes often omit the detailed proofs of stated theorems or just sketch out important ideas.

The book is divided into six parts. In the first part, we discuss the associated primes of ideals and, in particular, the persistence property and the stability index of these sets. In the second part, we investigate the asymptotic linearity of the Castelnuovo–Mumford regularity of powers of a homogeneous ideal. Most of our attention will be restricted to symbolic and ordinary powers of edge ideals of graphs. The third part of the book is devoted to the containments between symbolic and ordinary powers of ideals, focusing on squarefree monomial ideals and the defining ideals of schemes of fat points. In Part IV, we examine the very recently introduced notion of unexpected curves and the role of the SHGH conjecture in inspiring it. In Part V, we discuss the Waring problem for homogeneous polynomials. Specifically, we describe Sylvester's algorithm for binary forms and the connection to Strassen's conjecture. Part VI of the book is a summary of materials presented at the PRAGMATIC school. In particular, we have included a chapter on "The Art of Research," which aims at helping young researchers with advice on how to start a research project, on how to collaborate, on how to write up their results, and on how to present their findings.

We assume that the interested reader is familiar with basic concepts from commutative algebra. Unexplained notations and terminology can be found in standard texts [14, 25, 47, 63, 131, 137, 155, 166].

Torino, Italy   Enrico Carlini
New Orleans, LA, USA   Huy Tài Hà
Lincoln, NE, USA   Brian Harbourne
Hamilton, ON, Canada   Adam Van Tuyl
July 2017–December 2019

Acknowledgements

We begin by thanking the organizers of PRAGMATIC, namely Elena Guardo, Alfio Ragusa, Francesco Russo, and Giuseppe Zappalà, for the invitation to participate and for the wonderful experience. In addition, we are grateful for the referee's invaluable comments and suggestions, which helped improve this manuscript.

We also thank the following groups whose support made PRAGMATIC possible:

- Università di Catania (UNICT),
- Finanziamento della Ricerca d'Ateneo, Università di Catania (FIR UNICT),
- Gruppo Nazionale per le Strutture Algebriche, Geometriche e le loro Applicazioni (GNSAGA),
- Foundation Compositio Mathematica,
- Dipartimento di Matematica e Informatica, Università di Catania (DMI UNICT), and
- Scuola Superiore di Catania.

In addition, Enrico Carlini would like to thank the Politecnico di Torino—Finanziamento Diffuso, Tài Hà and Brian Harbourne were (during the school and the writing of this book) partially supported by a Simons Foundation Collaborative Grant, and Adam Van Tuyl was supported in part by an NSERC Discovery Grant.

Participants of PRAGMATIC 2017

Lecturers and Collaborators

Enrico Carlini
Politecnico di Torino (Italy)
enrico.carlini@polito.it

Brian Harbourne
University of Nebraska (USA)
brianharbourne@unl.edu

Huy Tài Hà
Tulane University (USA)
tha@tulane.edu

Adam Van Tuyl
McMaster University (Canada)
vantuyl@math.mcmaster.ca

PRAGMATIC 2017 participants

Organizers

Elena Guardo
Università di Catania (Italy)
guardo@dmi.unict.it

Alfio Ragusa
Università di Catania (Italy)
ragusa@dmi.unict.it

Francesco Russo
Università di Catania (Italy)
frusso@dmi.unict.it

Giuseppe Zappalà
Università di Catania (Italy)
zappalag@dmi.unict.it

Students

Iman Bahmani Jafarloo
Politecnico di Torino (Italy)
ibahmani@unito.it

Erin Bela
University of Notre Dame (USA)
ebela@nd.edu

Laura Brustenga Moncusi
Universitat Autónoma de Barcelona
(Spain)
laurea987@gmail.com

Yairon Cid
University of Barcelona (Spain)

ycid@ub.edu

Michela Di Marca
Università di Genova (Italy)
dimarca@dima.unige.it

Benjamin Drabkin
University of Nebraska (USA)
benjamin.drabkin@huskers.unl.edu

Maryam Ehya
Dalhousie University (Canada)
mr390045@dal.ca

Łucja Farnik
Polish Academy of Sciences (Poland)
lucja.farnik@gmail.com

Giuseppe Favacchio
Università di Catania (Italy)
favacchio@dmi.unict.it

Francesco Galuppi
Università di Ferrara (Italy)
glpfnc@unife.it

Lorenzo Guerrieri
Università di Catania (Italy)
guerrieri@dmi.unict.it

Sepehr Jafari
Università di Genova (Italy)
sepehr@dima.unige.it

Grzegorz Malara
Pedagogical University in Cracow
(Poland)
grzegorzmalara@gmail.com

Navid Nemati
Université Pierre et Marie Curie
(France)
navid.nemati@imj-prg.fr

Jonathan O'Rourke
Tulane University (USA)
jorourk2@tulane.edu

Giancarlo Rinaldo
Università di Trento (Italy)
giancarlo.rinaldo@unitn.it

William Trok
University of Kentucky (USA)

william.trok@uky.edu

Giuseppe Zito
Università di Catania (Italy)
giuseppezito@hotmail.it

Shreedevi K. Masuti
Università di Genova (Italy)

masuti@dima.unige.it

Alessandro Oneto
Inria Sophia Antipolis
Méditerranée (France)
alessandro.oneto@inria.fr

Beatrice Picone
Università di Catania (Italy)
beatrixpico@hotmail.it

Luca Sodomaco
Università di Florence (Italy)
lucasodomaco@gmail.com

Tran Thi Hieu Nghia
National University of Ireland (Ireland)
n.tranthihieu1@nuigalway.ie

Contents

Part I
Associated Primes of Powers of Ideals

Chapter 1
Associated Primes of Powers of Ideals

The primary decomposition of ideals in Noetherian rings is a fundamental result in commutative algebra and algebraic geometry. It is a far reaching generalization of the fact that every positive integer has a unique factorization into primes. We recall one version of this result.

Theorem 1.1 *Every ideal I in a Noetherian ring R has a minimal primary decomposition*

$$I = Q_1 \cap \cdots \cap Q_s$$

where each Q_i is a primary ideal and $Q_1 \cap \cdots \cap \widehat{Q}_j \cap \cdots Q_s \not\subset Q_j$ for all $j = 1, \ldots, s$. Furthermore, the set of associated primes *of I, that is,*

$$\mathrm{ass}(I) = \left\{ \sqrt{Q_1} = P_1, \ldots, \sqrt{Q_s} = P_s \right\}$$

is uniquely determined by I.

The primary decomposition of an ideal is a standard topic in most introductory commutative algebra books, e.g. see Atiyah-MacDonald [5, Chapter 5].

Starting in the 1970s, the following problem was investigated:

Question 1.2 Given an ideal I in a Noetherian ring R, how do the sets $\mathrm{ass}(I^s)$ change as s varies?

At first glance, it might be surprising that the set of associated primes of an ideal changes when you take its power. However, as the next example shows (and the many examples in the next chapter), new associated primes can appear in higher powers, and in fact, all sorts of pathological behaviour can occur.

© The Editor(s) (if applicable) and The Author(s), under exclusive licence to Springer Nature Switzerland AG 2020
E. Carlini et al., *Ideals of Powers and Powers of Ideals*, Lecture Notes of the Unione Matematica Italiana 27, https://doi.org/10.1007/978-3-030-45247-6_1

Example 1.3 In the ring $R = \mathbb{K}[x, y, z]$, consider the monomial ideal $I = \langle xy, xz, yz \rangle$. Then this ideal has the primary decomposition

$$I = \langle x, y \rangle \cap \langle x, z \rangle \cap \langle y, z \rangle = P_1 \cap P_2 \cap P_3.$$

On the other hand, the primary decomposition of I^2 is given by

$$I^2 = \langle x^2 y^2, x^2 yz, xy^2 z, x^2 z^2, xyz^2, y^2 z^2 \rangle$$
$$= \langle x, y \rangle^2 \cap \langle x, z \rangle^2 \cap \langle y, z \rangle^2 \cap \langle x^2, y^2, z^2 \rangle.$$

We thus have

$$\mathrm{ass}(I) = \{P_1, P_2, P_3\} \subsetneq \mathrm{ass}(I^2) = \mathrm{ass}(I) \cup \{\langle x, y, z \rangle\}.$$

In 1979, Brodmann [24] proved the following result which gives an asymptotic answer to Question 1.2.

Theorem 1.4 ([24]) *For any ideal $I \subseteq R$ in a Noetherian ring, there exists an integer s_0 such that*

$$\mathrm{ass}(I^s) = \mathrm{ass}(I^{s_0}) \text{ for all } s \geq s_0.$$

As we shall see in Chap. 2, Theorem 1.4 inspires a number of new questions, many of which only have partial solutions. Given the importance of Brodmann's result, the goal of this chapter is to sketch out the main ideas behind the proof of Theorem 1.4. A by-product of our approach is to learn some techniques related to associated primes that will hopefully be useful in your own research. As we move forward, R will always denote a Noetherian ring.

As a final comment, this chapter is greatly indebted not only to Brodmann's original paper, but to the monograph of McAdam [136] and the lecture notes of Swanson [159].

1.1 Associated Primes of Modules

We begin with a review/introduction to associated primes of modules. As we shall see, this is the correct point-of-view to take when studying the associated primes of I^s. Much of this material is standard. We use Villarreal's book [166] as a reference, although other books contain this material.

Definition 1.5 Let M be an R-module. For any $m \in M$, the *annihilator* of m is

$$\mathrm{ann}(m) = (0_M :_R m) = \{r \in R \mid rm = 0_M\}.$$

It is a straightforward exercise to show that $\mathrm{ann}(m)$ is an ideal of R.

Definition 1.6 Let $N \subseteq M$ be modules over R. A prime ideal $P \subseteq R$ is an *associated prime* of the R-module M/N if there exists some $m \in M$ such that

$$P = \mathrm{ann}(\overline{m}) = \{r \in R \mid r\overline{m} = 0_{M/N}\}$$

where $\overline{m}$ denotes the equivalence class of m in M/N, or equivalently,

$$P = (N :_R m) = \{r \in R \mid rm \in N\}.$$

Note that if $N = (0_M)$ is the zero submodule of M, then P is an associated prime of $M \cong M/(0_M)$ if there exists an $m \in M$ such that $(0_M :_R m) = P$. Thus P is an associated prime of the module M if and only if $P = \mathrm{ann}(m)$ for some $m \in M$.

Definition 1.7 Let $N \subseteq M$ be modules over R. The *set of associated primes* of M/N is

$$\mathrm{Ass}_R(M/N) = \{P \subseteq R \mid P \text{ a prime ideal associated to } M/N\}.$$

We now state some useful facts about $\mathrm{Ass}_R(M/N)$.

Theorem 1.8 ([166, Corollary 2.1.18]) *Let $N \subseteq M$ be modules over a Noetherian ring R. Then*

$$|\mathrm{Ass}_R(M/N)| < \infty.$$

Notice that we are using a slightly different notation for the set of associated primes for modules versus the set of associated primes of an ideal (as in Theorem 1.1). However, the relationship is explained in the next theorem.

Theorem 1.9 ([166, Corollary 2.1.28]) *For any ideal $J \subseteq R$,*

$$\mathrm{Ass}_R(R/J) = \mathrm{ass}(J).$$

1.2 Reducing the Problem

The strategy behind the proof of Theorem 1.4 is to focus on the set of associated primes of the R-module I^s/I^{s+1}. We now explain why this is the case.

The reduction of the problem comes from the fact that we have the following containments of sets.

Lemma 1.10 *For any ideal $I \subseteq R$ and any integer $s \geq 1$, we have*

$$\mathrm{Ass}_R(I^s/I^{s+1}) \subseteq \mathrm{Ass}_R(R/I^{s+1}) \subseteq \mathrm{Ass}_R(I^s/I^{s+1}) \cup \mathrm{Ass}_R(R/I^s).$$

Proof The proof of this fact exploits the natural short exact sequence

$$0 \longrightarrow I^s/I^{s+1} \xrightarrow{f} R/I^{s+1} \xrightarrow{g} R/I^s \longrightarrow 0$$

where f is the identity map, and g is the ring homomorphism $x + I^{s+1} \mapsto x + I^s$.

Suppose $P \in \mathrm{Ass}_R(I^s/I^{s+1})$, i.e., there exits some $\overline{m} = m + I^{s+1} \in I^s/I^{s+1}$ such that $P = \mathrm{ann}(\overline{m})$. But then $f(\overline{m}) = \overline{m} \in R/I^{s+1}$, and so P is also an associated prime of R/I^{s+1}. This proves $\mathrm{Ass}_R(I^s/I^{s+1}) \subseteq \mathrm{Ass}_R(R/I^{s+1})$.

Now suppose that $P \in \mathrm{Ass}_R(R/I^{s+1}) \setminus \mathrm{Ass}_R(I^s/I^{s+1})$ with $P = \mathrm{ann}(\overline{m})$ for some $\overline{m} \in R/I^{s+1}$. Note that $\overline{m} \notin I^s/I^{s+1}$, because if it was, then P would also be an associated prime of I^s/I^{s+1}. So $g(\overline{m}) = m + I^s \neq 0_{R/I^s}$. To finish the proof, we will show that $P = \mathrm{ann}(\overline{m}) = \mathrm{ann}(g(\overline{m}))$, i.e., $P \in \mathrm{Ass}_R(R/I^s)$.

If $r \in \mathrm{ann}(\overline{m})$, then $rm \in I^{s+1} \subseteq I^s$, and thus $r \in \mathrm{ann}(g(\overline{m}))$. To prove the reverse containment, suppose that there exits some $t \in \mathrm{ann}(g(\overline{m})) \setminus \mathrm{ann}(\overline{m})$. In particular, this means that $tm \in I^s$, but $tm \notin I^{s+1}$. Hence $tm + I^{s+1} \in I^s/I^{s+1}$ is a non-zero element.

We now claim that $\mathrm{ann}(\overline{tm}) = \mathrm{ann}(\overline{m})$, where we view $\overline{tm} = tm + I^{s+1}$ as an element of I^s/I^{s+1} and $\overline{m} = m + I^{s+1}$ as an element of R/I^{s+1}. The containment $\mathrm{ann}(\overline{m}) \subseteq \mathrm{ann}(\overline{tm})$ is straightforward because if $rm \in I^{s+1}$, then $r(tm) \in I^{s+1}$. Now consider $r \in \mathrm{ann}(\overline{tm})$. Then $rt \in \mathrm{ann}(\overline{m}) = P$. The element t cannot be in P. Indeed, if $t \in P$, then $tm \in I^{s+1}$, which contradicts our earlier fact that $tm \notin I^{s+1}$. So, since P is prime and $t \notin P$, we have $r \in P = \mathrm{ann}(\overline{m})$. But then $P = \mathrm{ann}(\overline{tm})$ is an associated prime of I^s/I^{s+1} which contradicts our choice of P. So, no such t can exist, i.e., $\mathrm{ann}(g(\overline{m})) \setminus \mathrm{ann}(\overline{m}) = \emptyset$, as desired.

Brodmann's proof reduces to proving the following theorem.

Theorem 1.11 *For any ideal $I \subseteq R$, there exists an integer s^* such that*

$$\mathrm{Ass}_R(I^s/I^{s+1}) = \mathrm{Ass}_R(I^{s^*}/I^{s^*+1}) \text{ for all } s \geq s^*.$$

Indeed, we can use the above statement to prove Brodmann's result.

Proof We want to show that there exists an integer s_0 such that for all $s \geq s_0$,

$$\mathrm{ass}(I^s) = \mathrm{Ass}_R(R/I^s) = \mathrm{Ass}_R(R/I^{s+1}) = \mathrm{ass}(I^{s+1}).$$

By Theorem 1.11, there exists an integer s^* such that if $s \geq s^*$,

$$\mathrm{Ass}_R(I^{s+1}/I^{s+2}) = \mathrm{Ass}_R(I^s/I^{s+1}) \subseteq \mathrm{Ass}_R(R/I^{s+1})$$

where the last containment follows from Lemma 1.10. Using this inclusion, and again using Lemma 1.10, we have the following inclusions:

$$\mathrm{Ass}_R(R/I^{s+2}) \subseteq \mathrm{Ass}_R(R/I^{s+1}) \cup \mathrm{Ass}_R(I^{s+1}/I^{s+2}) \subseteq \mathrm{Ass}_R(R/I^{s+1}).$$

It then follows that for all $t \geq 1$,

$$\cdots \subseteq \mathrm{Ass}_R(R/I^{s+t}) \subseteq \mathrm{Ass}_R(R/I^{s+t-1}) \subseteq \cdots \subseteq \mathrm{Ass}_R(R/I^{s+1}).$$

By Theorem 1.8 we know that $|\mathrm{Ass}_R(R/I^{s+1})| < \infty$, so this sequence must eventually stabilize. That is, there exists some $s_0 \geq s^\star$ such that for all $s \geq s_0$, we have

$$\mathrm{Ass}_R(R/I^s) = \mathrm{Ass}_R(R/I^{s+1}).$$

In other words, the sets $\mathrm{ass}(I^s)$ stabilize for $s \geq s_0$.

Remark 1.12 Note that the $s^\star$ of Theorem 1.11 is not necessarily the same s_0 of Theorem 1.4. However, as we will see in the next chapter (see Lemma 2.5) that when I is a monomial ideal, these values are the same.

1.3   Sketch of the Missing Details

The discussion in the previous section seems to imply that Brodmann's proof is not overly complicated. However, the "nitty-gritty" details of Theorem 1.4 are embedded in the proof of Theorem 1.11. We now attempt to explain the main steps one needs to prove Theorem 1.11.

The first step is to change the "point-of-view" again. Instead of viewing I^s/I^{s+1} as an R-module, you want to view it as an R/I-module. In particular, one needs to show:

Lemma 1.13 *Let $I \subseteq R$ be an ideal and s a positive integer. Then*

$$\mathrm{Ass}_R(I^s/I^{s+1}) = \mathrm{Ass}_R(I^{s+1}/I^{s+2}) \ \text{ if and only if } \ \mathrm{Ass}_{R/I}(I^s/I^{s+1}) = \mathrm{Ass}_{R/I}(I^{s+1}/I^{s+2}).$$

Proof This result will follow from the fact that

$$P \in \mathrm{Ass}_R(I^s/I^{s+1}) \ \text{ if and only if } \ P/I \in \mathrm{Ass}_{R/I}(I^s/I^{s+1}).$$

Suppose $P \in \mathrm{Ass}_R(I^s/I^{s+1})$. So, there exists an $m + I^{s+1} \in I^s/I^{s+1}$ such that $P = \{r \in R \mid r(m + I^{s+1}) = 0 + I^{s+1}\}$. But then

$$\begin{aligned} P/I &= \{r + I \mid r \in P\} \\ &= \{r + I \mid r(m + I^{s+1}) = 0 + I^{s+1}\} \\ &= \{r + I \mid (r + I)(m + I^{s+1}) = 0 + I^{s+1}\} = (0 :_{R/I} m + I^{s+1}). \end{aligned}$$

In other words, $P/I \in \mathrm{Ass}_{R/I}(I^s/I^{s+1})$.

Conversely, if $P/I \in \text{Ass}_{R/I}(I^s/I^{s+1})$, then

$$P/I = \{r + I \mid (r + I)(m + I^{s+1}) = 0 + I^{s+1}\}$$

for some $m + I^{s+1}$ in I^s/I^{s+1}. Since $P = \{r \mid r + I \in P/I\}$,

$$P = \{r \mid (r+I)(m+I^{s+1}) = rm+I^{s+1} = 0+I^{s+1}\} = \{r \mid r(m+I^{s+1}) = 0+I^{s+1}\}.$$

Thus, $P = (0 :_R m + I^{s+1})$, and thus $P \in \text{Ass}_R(I^s/I^{s+1})$.

Thanks to Lemma 1.13 it is enough to prove Theorem 1.11 for $\text{Ass}_{R/I}(I^s/I^{s+1})$. The second step is to work in a new ring constructed from I and R.

Definition 1.14 Given an ideal I in the ring R, the *associated graded ring* is the ring

$$G_I(R) = \bigoplus_{s=0}^{\infty} \frac{I^s}{I^{s+1}} \text{ where } I^0 = R.$$

The ring $G_I(R)$ is a graded ring where the d-th graded piece is $[G_I(R)]_d = I^d/I^{d+1}$. In particular, the 0-th graded piece is $[G_I(R)]_0 = R/I$. In this ring the multiplication of homogeneous elements is defined by

$$I^d/I^{d+1} \times I^e/I^{e+1} \xrightarrow{\times} I^{d+e}/I^{d+e+1}$$
$$(F + I^{d+1}, G + I^{e+1}) \mapsto (FG + I^{d+e+1}).$$

Note that you need to verify that this map is well-defined, i.e., it is independent of your coset representatives F and G.

The following fact about $G_I(R)$ is then used in Corollary 1.18 below; it can be found in Atiyah-MacDonald's book [5].

Theorem 1.15 ([5, Proposition 10.22]) *If R is a Noetherian ring, and $I \subseteq R$ is an ideal, then $G_I(R)$ is a Noetherian ring.*

The third step is to relate the prime ideals that appear in $\text{Ass}_{R/I}(I^s/I^{s+1})$ to the prime ideals of the ring $G_I(R)$. The desired relationship is described as a corollary to the next theorem which can be found in Swanson's paper [159].

Theorem 1.16 ([159, Proposition 5.7]) *Let G be a submonoid of $\mathbb{N}^n$. Let R be a G-graded ring and M a G-graded R-module. Suppose that $P \in \text{Ass}_{R_0}(M_g)$, where M_g is the degree g piece of M with $g \in G$. Then there exits a prime ideal $Q \in \text{Ass}_R(M)$ such that $Q \cap R_0 = P$. Furthermore, $Q \in \text{Ass}_R(R)$.*

Corollary 1.17 *Let $I \subseteq R$ be an ideal, and suppose that*

$$\wp \in \bigcup_{s \geq 0} \mathrm{Ass}_{R/I}(I^s/I^{s+1}).$$

Then there exists a prime ideal $\wp^\star \subseteq G_I(R)$ such that

(i) $\wp^\star \cap [G_I(R)]_0 = \wp^\star \cap (R/I) = \wp$.

(ii) $\wp^\star \in \mathrm{Ass}_{G_I(R)}(G_I(R))$.

Proof One applies Theorem 1.16 to our specific case. In particular, to apply this theorem, let $G = \mathbb{N}$, let M and R be our ring $G_I(R)$, and let $g = s$.

Corollary 1.17 then gives the following corollary.

Corollary 1.18 *Let $I \subseteq R$ be an ideal. Then*

$$\left| \bigcup_{s \geq 0} \mathrm{Ass}_{R/I}(I^s/I^{s+1}) \right| < \infty.$$

Proof Corollary 1.17 (i) implies that each distinct prime $\wp \in \bigcup_{s \geq 0} \mathrm{Ass}_{R/I}(I^s/I^{s+1})$ gives rise to a distinct prime $\wp^\star$. So, if $\left| \bigcup_{s \geq 0} \mathrm{Ass}_{R/I}(I^s/I^{s+1}) \right| = \infty$, then Corollary 1.17 (ii) would imply that $|\mathrm{Ass}_{G_I(R)}(G_I(R))| = \infty$. But because $G_I(R)$ is Noetherian by Theorem 1.15, it follows from Theorem 1.8 that $|\mathrm{Ass}_{G_I(R)}(G_I(R))| < \infty$, thus giving a contradiction.

There are two technical arguments that need to be made. The first proof follows McAdam's proof [136, Lemma 1.1].

Lemma 1.19 *For any ideal $I \subseteq R$, there exists an integer ℓ such that for all $s \geq \ell$,*

$$\langle 0_{G_I(R)} :_{G_I(R)} [G_I(R)]_1 \rangle \cap [G_I(R)]_s = (0_{G_I(R)}).$$

Proof The ideal $\langle 0_{G_I(R)} :_{G_I(R)} [G_I(R)]_1 \rangle$ is a homogeneous ideal in the Noetherian graded ring $G_I(R)$. Suppose that $F_1, \ldots, F_s$ are the generators of this ideal. Set $\ell = 1 + \max\{\deg(F_1), \ldots, \deg(F_s)\}$. Now suppose that F is a homogeneous element of degree $s \geq \ell$ in this ideal, i.e., $F = H_1 F_1 + \cdots + H_s F_s$ with each H_i homogeneous. Moreover we must have $\deg H_i \geq 1$ for all i. But this means that $H_i \in I/I^2 G_I(R)$, so $H_i F_i = 0_{G_I(R)}$, and thus $F = 0_{G_I(R)}$.

In other words, all the elements in $G_I(R)$ that annihilate the degree one piece $[G_I(R)]_1 = I/I^2$ have degree less than ℓ. This lemma is then used to prove the next lemma.

Lemma 1.20 *For any ideal $I \subseteq R$, there exists an integer ℓ such that for all $s \geq \ell$,*

$$\mathrm{Ass}_{R/I}(I^s/I^{s+1}) \subseteq \mathrm{Ass}_{R/I}(I^{s+1}/I^{s+2}).$$

Proof Let $G = G_I(R)$, and let ℓ be as in Lemma 1.19. Let $s \geq \ell$, and $\wp \in \mathrm{Ass}_{R/I}(I^s/I^{s+1}) = \mathrm{Ass}_{G_0}(G_s)$. So, there exits a homogeneous element $c + I^{s+1}$ of degree s in G_s such that

$$\wp = \langle 0 + I^{s+1} :_{G_0} c + I^{s+1} \rangle.$$

We now claim that $\wp = \langle 0 + I^{s+2} :_{G_0} (c + I^{s+1})I/I^2 \rangle$ where $(c + I^{s+1})I/I^2$ is a submodule of I^{s+1}/I^{s+2}.

The containment $\wp = \langle 0 + I^{s+1} :_{G_0} c + I^{s+1} \rangle \subseteq \langle 0 + I^{s+2} :_{G_0} (c + I^{s+1})I/I^2 \rangle$ is immediate, so it suffices to verify the reverse containment. Suppose that $r + I \in \langle 0 + I^{s+2} :_{G_0} (c + I^{s+1})I/I^2 \rangle$. So $(rc + I^{s+1})I/I^2 = 0 + I^{s+2}$. But this means that

$$rc + I^{s+1} \in (0 + I^{s+2} :_{G_0} I/I^2) \cap I^s/I^{s+1}$$

$$\subseteq (0_{G_I(R)} :_{G_I(R)} [G_I(R)]_1) \cap [G_I(R)]_s.$$

By Lemma 1.19, this means that $rc + I^{s+1} = 0 + I^{s+1}$. Thus $r + I \in \langle 0 + I^{s+1} :_{G_0} c + I^{s+1} \rangle = \wp$, as desired.

We now show that $\wp \in \mathrm{Ass}_{R/I}(I^{s+1}/I^{s+2})$. We can localize so that we can assume that $\wp$ is a maximal ideal of R/I. We first claim that there is an $r + I^2 \in I/I^2$ such that

$$(r + I^2)(c + I^{s+1}) \neq 0 + I^{s+2}.$$

Indeed, if there was no such r, then we would have

$$\wp = \langle 0 + I^{s+1} :_{G_0} (c + I^{s+1})I/I^2 \rangle = R/I \neq \wp.$$

So, let $r + I^2$ be such an element. Then

$$\wp = \langle 0 + I^{s+2} :_{G_0} (c + I^{s+1})I/I^2 \rangle \subseteq \langle 0 + I^{s+2} :_{G_0} (cr + I^{s+2}) \rangle \subsetneq R/I.$$

Because $\wp$ is a maximal ideal, we must have $\wp = \langle 0 + I^{s+2} :_{G_0} (cr + I^{s+2}) \rangle$, i.e., $\wp \in \mathrm{Ass}_{R/I}(I^{s+1}/I^{s+2})$, as desired. $\qquad \blacksquare$

We can use these pieces to prove Theorem 1.11.

Proof By Lemma 1.13, it is enough to show that there exists an integer s_0 such that

$$\mathrm{Ass}_{R/I}(I^s/I^{s+1}) = \mathrm{Ass}_{R/I}(I^{s_0}/I^{s_0+1}) \text{ for all } s \geq s_0.$$

It follows from Lemma 1.20 that there exists an integer ℓ such that

$$\operatorname{Ass}_{R/I}(I^{\ell}/I^{\ell+1}) \subseteq \operatorname{Ass}_{R/I}(I^{\ell+1}/I^{\ell+2}) \subseteq \cdots \subseteq \bigcup_{s \geq 0} \operatorname{Ass}_{R/I}(I^s/I^{s+1}).$$

But by Corollary 1.18, $\left|\bigcup_{s \geq 0} \operatorname{Ass}_{R/I}(I^s/I^{s+1})\right| < \infty$. We thus have a sequence of subsets in a finite set, where the i-th set is contained in the $(i+1)$-th set. So, there must exist some s_0 such that

$$\operatorname{Ass}_{R/I}(I^{s_0}/I^{s_0+1}) = \operatorname{Ass}_{R/I}(I^{s_0+1}/I^{s_0+2}) = \cdots$$

thus completing the proof.

1.4 Final Comments

Brodmann's Theorem (Theorem 1.4) is a good example of the idea in commutative algebra that ideals behave "nicely" asymptotically (see also the later chapters on the powers of ideals and regularity, giving more evidence of this idea).

Of course, Brodmann's Theorem also inspires a number of natural questions (e.g., given an ideal I, can we determine the value of s_0). In the next chapter we will explore some of these problems in the case I is a monomial ideal.

We end with a recent result of Hà, Nguyen, Trung, and Trung that shows if $s < s_0$, the sets $\operatorname{ass}(I^s)$ need not be related to each other. Moreover, we can make examples where s_0 is arbitrarily large (although we may need to work in a very large polynomial ring!).

Theorem 1.21 ([95, Corollary 6.8]) *Let Γ be any finite subset of $\mathbb{N}^+$. Then there exists a monomial ideal I in a polynomial ring R such that*

$$\mathfrak{m} \in \operatorname{ass}(I^s) \text{ if and only if } s \in \Gamma.$$

Here, $\mathfrak{m}$ is the unique maximal monomial ideal of R.

Remark 1.22 The above result answers an old question first raised by Ratliff [148].

Chapter 2
Associated Primes of Powers of Squarefree Monomial Ideals

In the previous chapter, we looked at a result of Brodmann (Theorem 1.4) concerning the associated primes of powers of ideals. This theorem inspires a number of natural questions. To state these questions, we introduce some suitable terminology.

Definition 2.1 The *index of stability* of an ideal I in a Noetherian ring R, denoted $\mathrm{astab}(I)$, is defined to be

$$\mathrm{astab}(I) := \min\{s_0 \mid \mathrm{ass}(I^s) = \mathrm{ass}(I^{s_0}) \text{ for all } s \geq s_0\}.$$

Definition 2.2 An ideal I in a Noetherian ring R is said to have the *persistence property* if $\mathrm{ass}(I^i) \subseteq \mathrm{ass}(I^{i+1})$ for all $i \geq 1$.

Brodmann's result is the inspiration for the following questions:

Question 2.3 Let I be an ideal of a Noetherian ring R.

 (*i*) What is $\mathrm{astab}(I)$?
 (*ii*) Does I have the persistence property?
 (*iii*) What are the elements of $\mathrm{ass}(I^s)$ with $s \geq \mathrm{astab}(I)$?

In general, these questions appear to be quite difficult. (Note that Theorem 1.21 implies the existence of ideals that fail the persistence property.) In this chapter, we want to focus on the case that I is a (squarefree) monomial ideal in a polynomial ring $R = \mathbb{K}[x_1, \ldots, x_n]$. In this context, we have a much better understanding of the problems raised in Question 2.3.

E. Carlini et al., *Ideals of Powers and Powers of Ideals*, Lecture Notes of the Unione Matematica Italiana 27, https://doi.org/10.1007/978-3-030-45247-6_2

2.1   General (Useful) Facts About Monomial Ideals

As mentioned above, we are going to focus on the case of monomial ideals. This tighter focus imposes restrictions on what primes can be associated primes, and it gives us some information about the annihilator.

Lemma 2.4 *Let I be any monomial ideal of $R = \mathbb{K}[x_1, \ldots, x_n]$.*

(i) *If $P \in \mathrm{ass}(I)$, then P is also a monomial ideal, that is, $P = \langle x_{i_1}, \ldots, x_{i_r} \rangle$ for some $\{x_{i_1}, \ldots, x_{i_r}\} \subseteq \{x_1, \ldots, x_n\}$.*

(ii) *If $P \in \mathrm{ass}(I)$, then there exists a monomial $m \in R \setminus I$ such that $I : \langle m \rangle = P$.*

Proof For (i), suppose that I has a monomial generator m that does not have the form x_i^a for some variable x_i. We can then factor m as $m = m_1 m_2$ where $\gcd(m_1, m_2) = 1$ with $m_1 \neq 1$ and $m_2 \neq 1$. We then have the identity

$$I = (J + \langle m_1 \rangle) \cap (J + \langle m_2 \rangle)$$

where J is the monomial ideal generated by monomial generators of I except m. By repeatedly applying this identity, we can rewrite I as the intersection of ideals of the form $\langle x_{i_1}^{a_{i_1}}, \ldots, x_{i_r}^{a_{i_r}} \rangle$. Each of these ideals are $\langle x_{i_1}, \ldots, x_{i_r} \rangle$-primary. Because the associated primes of an ideal I are uniquely determined by I, any associated prime of I must have the form $P = \langle x_{i_1}, \ldots, x_{i_r} \rangle$ for some $\{x_{i_1}, \ldots, x_{i_r}\} \subseteq \{x_1, \ldots, x_n\}$.

For statement (ii), since $P \in \mathrm{ass}(I)$, there exists an $f \in R$ such that $I : \langle f \rangle = P$. If f is not a monomial, we can write it as $f = c_1 m_1 + \cdots + c_s m_s$ with $c_i \in k$ and m_i a monomial. By (i), we know that $P = \langle x_{i_1}, \ldots, x_{i_r} \rangle$. So, for any $x_j \in P$,

$$f x_j = c_1 m_1 x_j + \cdots + c_s m_s x_j \in I \Rightarrow m_k x_j \in I \text{ for each } k \in \{1, \ldots, s\}$$

since I is a monomial ideal. But this means that $x_j \in I : \langle m_k \rangle$ for all $k \in \{1, \ldots, s\}$. Since this is true for each $x_j \in P$, we have

$$P \subseteq \bigcap_{i=1}^{s} I : \langle m_i \rangle.$$

If $g \in \bigcap_{i=1}^{s} I : \langle m_i \rangle$, then $fg = c_1 m_1 g + \cdots + c_s m_s g \in I$. This means that $g \in I : \langle f \rangle = P$.

We have thus shown that $P = \bigcap_{i=1}^{s} I : \langle m_i \rangle$. But because a prime ideal is an irreducible ideal, we must have $P = I : \langle m_i \rangle$ for some $i \in \{1, \ldots, s\}$.

In Brodmann's proof of the asymptotic stability of $\mathrm{ass}(I^s)$, he used the stability of $\mathrm{Ass}_R(I^s / I^{s+1})$ to prove the stability of $\mathrm{Ass}_R(R/I^{s+1})$. In general, these sets are not equal. However, in the case of monomial ideals, these sets are the same.

Lemma 2.5 *For any monomial ideal $I \subseteq R$,*

$$\mathrm{Ass}_R(I^s/I^{s+1}) = \mathrm{Ass}_R(R/I^{s+1}) \text{ for all } s \geq 0.$$

Proof By Lemma 1.10, we always have $\mathrm{Ass}_R(I^s/I^{s+1}) \subseteq \mathrm{Ass}_R(R/I^{s+1})$ for any ideal in a Noetherian ring. It suffices to prove the reverse containment for monomial ideals. Let $P \in \mathrm{Ass}_R(R/I^{s+1}) = \mathrm{ass}(I^{s+1})$. By Lemma 2.4 (i) and (ii) there exists a monomial $m \in R$ such that

$$P = \langle x_{i_1}, \ldots, x_{i_r} \rangle = I^{s+1} : \langle m \rangle \text{ with } m \in R \setminus I^{s+1}.$$

So, for any $x_j \in P$,

$$m x_j = m_1 \cdots m_{s+1} M \in I^{s+1}$$

with m_i a monomial generator of I and M a monomial. After relabelling, we can assume that $x_j \mid (m_{s+1}M)$. So, $m_1 \cdots m_s \mid m$, which implies that $m \in I^s$.

We thus have $m \in I^s \setminus I^{s+1}$ and $P = I^{s+1} : \langle m \rangle$. But this is precisely the condition for $P \in \mathrm{Ass}_R(I^s/I^{s+1})$.

Corollary 2.6 *For a monomial ideal $I \subseteq R = \mathbb{K}[x_0, \ldots, x_n]$,*

$$\mathrm{astab}(I) = \min \left\{ s_0 \; \middle| \; \mathrm{Ass}_R(I^s/I^{s+1}) = \mathrm{Ass}_R(I^{s_0}/I^{s_0+1}) \text{ for all } s \geq s_0 \right\}.$$

For any monomial ideal I, we let $G(I)$ denote the unique set of minimal generators of I. For any monomial $m = x_1^{a_1} \cdots x_n^{a_n}$, we define the support of m to be

$$\mathrm{supp}(m) = \{x_i \mid a_i > 0\}.$$

We end this section with a useful localization "trick". This theorem is useful because it sometimes allows us to reduce to the case that $P = \langle x_1, \ldots, x_n \rangle$ is the unique monomial ideal that is also a maximal ideal. We first state a lemma which describes the localization of a monomial ideal at an associated prime ideal.

Lemma 2.7 *Let $I \subseteq R = \mathbb{K}[x_1, \ldots, x_n]$ be a monomial ideal and suppose that $P = \langle x_{i_1}, \ldots, x_{i_r} \rangle \in \mathrm{Ass}_R(R/I)$. Then*

$$I_P = \langle m_1 \mid m = m_1 m_2 \in G(I) \rangle \subseteq R_P = \mathbb{K}[x_{i_1}, \ldots, x_{i_r}]$$

where m_1 is a monomial in the variables $W = \{x_{i_1}, \ldots, x_{i_r}\}$ and m_2 is a monomial in the variables $\{x_1, \ldots, x_n\} \setminus W$.

Proof This result follows from the fact localizing R, respectively I, at the prime ideal $P = \langle x_{i_1}, \ldots, x_{i_r} \rangle$ to form R_P, respectively I_P, is equivalent to setting all the variables in R, respectively I, not in P equal to one.

Theorem 2.8 *Let $I \subseteq R = \mathbb{K}[x_1, \ldots, x_n]$ be a monomial ideal. Then*

$$P = \langle x_{i_1}, \ldots, x_{i_r} \rangle \in \mathrm{Ass}_R(\mathbb{K}[x_1, \ldots, x_n]/I^s)$$

if and only if

$$P = \langle x_{i_1}, \ldots, x_{i_r} \rangle \in \mathrm{Ass}_S(\mathbb{K}[x_{i_1}, \ldots, x_{i_r}]/(I_P)^s),$$

where $S = \mathbb{K}[x_{i_1}, \ldots, x_{i_r}] = R_P$.

Proof After relabelling, we can assume that $P = \langle x_1, \ldots, x_m \rangle$, and we let $\mathbb{K}[P] = S$.

($\Rightarrow$) Suppose that $P = \langle x_1, \ldots, x_m \rangle \in \mathrm{Ass}_R(R/I^s)$. Then there exists a monomial m such that $I^s : \langle m \rangle = P$. We can rewrite m as $m = m_1 m_2$, where m_1 is a monomial in $\mathbb{K}[P]$ and m_2 is a monomial in $\{x_{m+1}, \ldots, x_n\}$.

For any monomial u in the variables $\{x_{m+1}, \ldots, x_n\}$, we claim that $I^s : \langle mu \rangle = I^s : \langle m \rangle$. To see this, first note that $mu \notin I^s$, for if it were, then $u \in I^s : \langle m \rangle = P$, which is false since $u \notin P$. For any $x_j \in P$, $(mu)x_j = (mx_j)u \in I^s$ since $mx_j \in I^s$. So $P \subseteq I^s : \langle mu \rangle$. Now take any monomial $w \in R$ such that $w \in I^s : \langle mu \rangle$. If w is a monomial only in the variables $\{x_{m+1}, \ldots, x_n\}$, then $(mu)w = m(uw) \in I^s$ implies that $uw \in P$, which is again a contradiction since neither u nor w is divisible by any of $\{x_1, \ldots, x_m\}$. So $I^s : \langle mu \rangle = P$.

As a consequence, we can multiply m by a suitable monomial u in the variables $\{x_{m+1}, \ldots, x_n\}$ so that $m = m_1 m_2$ with $m_2 = (x_{m+1} \cdots x_n)^s m_2'$. That is,

$$I^s : \langle m_1 (x_{m+1} \cdots x_n)^s m_2' \rangle = \langle x_1, \ldots, x_m \rangle.$$

We now show that $I_P^s : \langle m_1 \rangle = \langle x_1, \ldots, x_m \rangle$ in $\mathbb{K}[P]$. First, we show that $m_1 \notin I_P^s$. If it were, then there exists monomials $w_1, \ldots, w_s \in I_P$ such that $m_1 = w_1 \cdots w_s M$ for some monomial $M \in \mathbb{K}[P]$. But then

$$m = m_1 (x_{m+1} \cdots x_n)^s m_2' = (w_1 \cdots w_s M)(x_{m+1} \cdots x_n)^s m_2'$$

$$= [w_1 (x_{m+1} \cdots x_n)][w_2 (x_{m+1} \cdots x_n)] \cdots [w_s (x_{m+1} \cdots x_n)] M m_2'.$$

Note that for each $i = 1, \ldots, s$, $w_i (x_{m+1} \cdots x_n) \in I$. The above expression thus implies that $m \in I^s$, a contradiction. So $m_1 \notin I_P^s$.

For any $x_i \in P$, we have $m x_i \in I^s$. Thus, there exists $w_1, \ldots, w_s \in I$ such that

$$m x_i = m_1 m_2 x_i = w_1 \cdots w_s N = w_{1,1} w_{1,2} \cdots w_{s,1} w_{s,2} N,$$

where we have written each w_i as $w_i = w_{i,1} w_{i,2}$ with $w_{i,1}$ a monomial in $\mathbb{K}[P]$ and $w_{i,2}$ a monomial in the remaining variables. Thus, if we compare the monomials in $\mathbb{K}[P]$ on both sides of the above expression, we get $(w_{1,1} \cdots w_{s,1}) \mid m_1 x_i$. But each $w_{i,1}$ is a generator of I_P by Lemma 2.7. So $m_1 x_i \in (I_P)^s$. Thus the maximal ideal $\langle x_1, \ldots, x_m \rangle \subseteq (I_P)^s : \langle m_1 \rangle$. Since $m_1 \notin (I_P)^s$, we have $(I_P)^s : \langle m_1 \rangle = \langle x_1, \ldots, x_m \rangle$, as desired.

$(\Leftarrow)$ Suppose $P = \langle x_1, \ldots, x_m \rangle \in \mathrm{Ass}(\mathbb{K}[P]/(I_P)^s)$. Thus there exists a monomial $m \in \mathbb{K}[P]$ with $m \notin (I_P)^s$ such that $(I_P)^s : \langle m \rangle = P$. We will show that

$$I^s : \langle m(x_{m+1} \cdots x_n)^s \rangle = \langle x_1, \ldots, x_m \rangle.$$

We first note that $m(x_{m+1} \cdots x_n)^s \notin I^s$. If it were, then there exist $w_1, \ldots, w_s \in I$ such that $m(x_{m+1} \cdots x_n)^s = w_1 \cdots w_s M$. Rewriting each w_i as $w_i = w_{i,1} w_{i,2}$, where $w_{i,1}$ is a monomial in $\mathbb{K}[P]$, and $w_{i,2}$ is a monomial in the variables $\{x_{m+1}, \ldots, x_n\}$, we have $w_{1,1} \cdots w_{s,1} \mid m$. But each $w_{i,1}$ corresponds to a generator of I_P by Lemma 2.7, so $m \in (I_P)^s$, contradicting the fact that $1 \notin (I_P)^s : \langle m \rangle$.

Now let x_i be a generator of P. In the ring $\mathbb{K}[P]$, $m x_i = m_1 \cdots m_s M$ with $m_i \in I_P$ for each i. But then in R,

$$m(x_{m+1} \cdots x_n)^s x_i = [m_1(x_{m+1} \cdots x_n)] \cdots [m_s(x_{m+1} \cdots x_n)] M.$$

For each $i = 1, \ldots, s$, the monomial $m_i(x_{m+1} \cdots x_n) \in I$. Indeed, since m_i is a generator of I_P, there is a generator $w_i \in I$ such that $w_i = m_i v_i$ with v_i in the variables $\{x_{m+1}, \ldots, x_n\}$ by Lemma 2.7. Since w_i is squarefree, $w_i \mid m_i(x_{m+1} \cdots x_n)$, and thus $m_i(x_{m+1} \cdots x_n) \in I$. Hence $P \subseteq I^s : \langle m(x_{m+1} \cdots x_n)^s \rangle$. For the reverse inclusion, consider any monomial $w \in I^s : \langle m(x_{m+1} \cdots x_n)^s \rangle$. If there exists some variable $x_i \in \{x_{m+1}, \ldots, x_n\}$ such that $x_i \mid w$, then $\frac{w}{x_i} \in I^s : \langle m(x_{m+1} \cdots x_n)^s \rangle$. Indeed, if $w = w' x_i$, then $m(x_{m+1} \cdots x_n)^s (w' x_i) = m_1 \cdots m_s M$, and because each m_i is squarefree and can be divisible by at most one x_i in $\{x_{m+1}, \ldots, x_n\}$, we have $x_i \mid M$. So

$$m(x_{m+1} \cdots x_n)^s w' = m_1 \cdots m_s \left(\frac{M}{x_i} \right),$$

whence $w' \in I^s : \langle m(x_{m+1} \cdots x_n)^s \rangle$. We can now reduce to the case that w is a monomial only in the variables of $\{x_1, \ldots, x_m\}$. But this just means that $w \in P$, as desired.

Remark 2.9 The above result first appeared in a paper of Francisco, Hà, and Van Tuyl [76, Lemma 2.11], but was expressed in the language of induced subhypergraphs of hypergraphs.

2.2   Monomial Ideals and Connections to Graph Theory: A First Look

Monomial ideals, most notably squarefree monomial ideals, have strong connections to combinatorics (e.g., graph theory, hypergraphs, simplicial complexes). In fact, as we shall see throughout this monograph, a combinatorial point-of-view can give us some new insights into algebraic questions, and vice versa. In this section, we set up the dictionary between graph theory and squarefree monomial ideals. This dictionary gives us a language to describe results about associated primes in terms of the invariants and properties coming from graph theory. Future sections and chapters will build upon this correspondence.

We write $G = (V, E)$ to denote the *finite simple graph* with vertex set $V = \{x_1, \ldots, x_n\}$ and edge set E, i.e., E is a collection of subsets $e \subseteq V$ with $|e| = 2$. We may sometimes write $(V(G), E(G))$ if we wish to highlight the vertex set and edge set of G. By identifying the vertices of V with the variables of $R = \mathbb{K}[x_1, \ldots, x_n]$, we can construct two monomial ideals.

Definition 2.10   The *edge ideal* of G is the ideal

$$I(G) = \langle x_i x_j \mid \{x_i, x_j\} \in E \rangle$$

where the monomial generators correspond to the edges of G. The *cover ideal of G* is the ideal

$$J(G) = \bigcap_{\{x_i, x_j\} \in E} \langle x_i, x_j \rangle.$$

We pause to explain the significance of the name cover ideal.

Definition 2.11   A subset $W \subseteq V(G)$ is a *vertex cover* if $W \cap e \neq \emptyset$ for all $e \in E(G)$. A vertex cover W is a *minimal vertex cover* if no proper subset of W is a vertex cover.

As the next lemma shows, the generators of the cover ideal $J(G)$ correspond to the minimal vertex covers of G.

Lemma 2.12   *Let G be a graph with cover ideal $J(G)$. Then*

$$J(G) = \langle x^W \mid W \subseteq V(G) \text{ is a minimal vertex cover of } G \rangle$$

where $x^W := \prod_{x_i \in W} x_i$ if $W \subseteq V(G)$.

Proof   Let $L = \langle x^W \mid W \subseteq V(G) \text{ is a minimal vertex cover of } G \rangle$.

Let x^W be a generator of L with W a minimal vertex cover. Then, for every edge $e = \{x_i, x_j\} \in E(G)$, we have $W \cap e \neq \emptyset$. So, either $x_i \in W$ or $x_j \in W$. Consequently, either $x_i | x^W$ or $x_j | x^W$, whence $x^W \in \langle x_i, x_j \rangle$. Since e is arbitrary, we have

$$x^W \in \bigcap_{\{x_i, x_j\} \in E(G)} \langle x_i, x_j \rangle = J(G).$$

Conversely, let $m \in J(G)$ be any minimal generator. Note that m must be squarefree since $J(G)$ is the intersection of finitely many squarefree monomial ideals. So $m = x_{i_1} \cdots x_{i_r}$. Let $W = \{x_{i_1}, \ldots, x_{i_r}\}$. Since $m \in \langle x_i, x_j \rangle$ for each each $\{x_i, x_j\} \in E(G)$. either $x_i | m$ or $x_j | m$, and thus, $x_i \in W$ or $x_j \in W$. Thus W is a vertex cover. Let $W' \subseteq W$ be a minimal vertex cover. Because $x^{W'} \in L$ and $x^{W'}$ divides $m = x^W$, we have $m \in L$.

The minimal vertex covers of G are also related to the minimal primary decomposition of the edge ideal $I(G)$.

Lemma 2.13 *Let G be a graph with edge ideal $I(G)$. Then*

$$I(G) = \bigcap_{w \text{ is a minimal vertex cover of } G} \langle x \mid x \in W \rangle.$$

Proof Let K denote the ideal on the right hand side of the statement. If $x_i x_j \in I(G)$ is a generator, then for every minimal vertex cover W, either x_i or x_j is in W. Consequently, $x_i x_j \in \langle x \mid x \in W \rangle$ for all minimal vertex covers, thus proving the containment $I(G) \subseteq K$.

The ideal K is an intersection of squarefree monomial ideals, so K is a squarefree monomial ideal. Let $m \in K$ be a squarefree minimal generator of K. Suppose x_{i_1} divides m. The set $W \setminus \{x_{i_1}\}$ is a vertex cover, so there is some minimal vertex cover $W_1 \subseteq W \setminus \{x_{i_1}\}$. Since $m \in \langle x \mid x \in W_1 \rangle$, there is some $x_{i_2} \in W_1$ that divides m. If $\{x_{i_1}, x_{i_2}\} \in E(G)$, then $x_{i_1} x_{i_2} \in I(G)$ divides m. If $\{x_{i_1}, x_{i_2}\}$ is not an edge, then $W \setminus \{x_{i_1}, x_{i_2}\}$ is a vertex cover. Consequently, there is a minimal vertex cover $W_2 \subseteq W \setminus \{x_{i_1}, x_{i_2}\}$. Thus there is some $x_{i_3} \in W_2$ that divides m. If x_{i_3} is adjacent to x_{i_1} or x_{i_2}, then we have found a generator of $I(G)$ that divides m. Otherwise, $\{x_{i_1}, x_{i_2}, x_{i_3}\}$ is an independent set, and $W \setminus \{x_{i_1}, x_{i_2}, x_{i_3}\}$ is vertex cover.

Continuing in this fashion, we will eventually find a minimal vertex cover $W_t \subseteq W \setminus \{x_{i_1}, \ldots, x_{i_t}\}$ such that there is an $x_{i_{t+1}} \in W_t$ that divides m and is adjacent to one of $x_{i_1}, \ldots, x_{i_t}$ (in particular, when $\{x_{i_1}, \ldots, x_{i_t}\}$ is a maximal independent set). So, there is a generator of $I(G)$ that divides m. Thus $K \subseteq I(G)$.

We now review some relevant definitions from graph which describe either special constructions or special families of graphs. The *complement* of G is the graph $G^c = (V(G^c), E(G^c))$ where $V(G^c) = V(G)$ and edge set $E(G^c) = \{\{x_i, x_j\} \subseteq V(G) \mid \{x_i, x_j\} \notin E(G)\}$. In other words, it is the graph consisting of the non-edges of G. The *induced graph* of G on $W \subseteq V$ is the graph $G_W = (W, E(G_W))$, where $E(G_W) = \{\{x_i, x_j\} \in E(G) \mid \{x_i, x_j\} \subseteq W\}$.

The notion of an independent set is dual to vertex cover.

Fig. 2.1 The five cycle graph

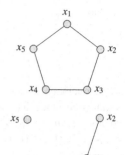

Fig. 2.2 The induced graph
$(C_5)_W$ with $W = \{x_2, x_3, x_5\}$

Definition 2.14 If G is a finite simple graph, then a subset $W \subseteq V(G)$ is an *independent set* if $V(G) \setminus W$ is a vertex cover. Equivalently, W is an independent set if the induced graph G_W has no edges. An independent set is a *maximal independent set* if it is maximal with respect to inclusion.

A *cycle* on n vertices, denoted C_n, is the graph

$$C_n = (\{x_1, \ldots, x_n\}, \{\{x_1, x_2\}, \{x_2, x_3\}, \ldots, \{x_{n-1}, x_n\}, \{x_n, x_1\}\}).$$

A graph G is a *perfect graph* if both G and G^c have no induced cycles C_n with n odd and $n \geq 5$. A *clique* on n vertices, denoted K_n, is the graph with $V(K_n) = \{x_1, \ldots, x_n\}$ and edge set $E(K_n) = \{\{x_i, x_j\} \mid 1 \leq i < j \leq n\}$. Finally, a graph is a *bipartite graph* if there is a partition of $V(G) = V_1 \cup V_2$ such that every $e \in E(G)$ has the property that $e \cap V_i \neq \emptyset$ for $i = 1, 2$.

A *colouring* of G is an assignment of colours to each vertex of G so that adjacent vertices receive distinct colours. Equivalently, a colouring of G is a partition of the vertices, say $V(G) = C_1 \cup \cdots \cup C_d$, where each C_i is an independent set. The *chromatic number* of G, denoted $\chi(G)$, is the minimum number of colours needed in a colouring.

Example 2.15 We illustrate some of these ideas with an example. The graph $G = C_5$ is given in Fig. 2.1. Then $\chi(G) = 3$ since we can colour vertices x_1, x_3 RED, vertices x_2, x_4 BLUE, and x_5 PURPLE, but there is no way to colour G with two colours. Note that $\{x_1, x_3\} \cup \{x_2, x_4\} \cup \{x_5\}$ is a partition of $V(G)$ into independent sets of C_5. When we consider $W = \{x_2, x_3, x_5\}$, the induced graph $G_W = (C_5)_W$ is the the graph in Fig. 2.2.

2.3   The Index of Stability

We now look at Question 2.3.(i) for monomial ideals. In general, we know very little about astab(I) for monomial ideals; this section summarizes some of the work in this area. One of the few general results about Question 2.3 is the following result of Hoa:

Theorem 2.16 ([114]) *Let I be a monomial ideal with n variables, s generators, and d the largest degree of a minimal generator. Then*

$$\text{astab}(I) \leq \max\left\{ d(ns + s + d))(\sqrt{n})^{n+1}(\sqrt{2}d)^{(n+1)(s-1)}, s(s+n)^4 s^{n+2} d^2 (2d^2)^{s^2-s+1} \right\}.$$

Example 2.17 The bound of Theorem 2.16 is very far from optimal. For example, for the ideal $I = \langle x_1 x_2, x_2 x_3 \rangle \subseteq \mathbb{K}[x_1, x_2, x_3]$, Theorem 2.16 gives the bound $\text{astab}(I) \leq 81,920,000$, but $\text{astab}(I) = 1$.

It would be nice to have better uniform bounds for all squarefree monomial ideals. J. Herzog has suggested that perhaps $\text{astab}(I) \leq n-1$, where n is the number of variables of R. In all known cases, this bounds appears to hold.

If we restrict to some families of monomial ideals related to finite simple graphs G. In particular, combinatorial information about G can now be used to place some bounds on $\text{astab}(I(G))$ and $\text{astab}(J(G))$, significantly improving upon the general bounds of Theorem 2.16. The proofs exploit many graph theoretic properties of the graphs G; because of the technical nature of the proofs, we refer the reader to the original papers by Simis, Vasconcelos, and Villarreal [153], Chen, Morey, and Sung [37], and Francisco, Hà, and Van Tuyl [76]

Theorem 2.18 *Let G be a finite simple graph.*

(i) [153, Theorem 5.9] *If G is a bipartite graph, then* $\text{astab}(I(G)) = 1$.

(ii) [37, Corollary 4.3] *If the smallest induced odd cycle of G has size $2k+1$, then* $\text{astab}(I(G)) \leq n - k$.

(iii) [76, Corollary 5.11] *If G is a perfect graph, then* $\text{astab}(J(G)) = \chi(G) - 1$.

Remark 2.19 Simis, Vasconcelos, and Villarreal's proof of statement (i) (see [153, Theorem 5.9]) actually shows that $I = I(G)$ is normally torsion free if G is bipartite, but this implies that $I^m = I^{(m)}$ for all $m \geq 1$. One can then show that $\text{ass}(I^m) = \text{ass}(I^{(m)}) = \text{ass}(I)$ for all $m \geq 1$.

2.4 Persistence of Primes

We now turn to Question 2.3.(ii), i.e., when does a monomial ideal have the persistence property. Persistence for monomial ideals fails in general. We do, however, have the following sufficient condition for persistence which is found in Martínez-Bernal, Morey, and Villarreal [135].

Theorem 2.20 *Suppose that I is a monomial ideal such that $I^{k+1} : I = I^k$ for all $k \geq 1$. Then I has the persistence property.*

Proof Let $P \in \text{ass}(I^k)$. By Theorem 2.8, we can assume that $P = \langle x_1, \ldots, x_n \rangle$ is the maximal homogeneous ideal; furthermore, it is enough to show that P persists.

Since $P \in \mathrm{ass}(I^k)$, there exists a monomial $m \in R \setminus I^k$ such that $I^k : \langle m \rangle = P$. Since $m \in R \setminus I^k$, $m \notin I^k = I^{k+1} : I$. So, there exists a monomial $q \in I$ such that $mq \notin I^{k+1}$. Now for each $i = 1, \ldots, n$, the variable x_i satisfies

$$(mq)x_i \in (mx_i)q \in I^k I = I^{k+1} \text{ because } mx_i \in I^k.$$

This implies that $P \subseteq I^{k+1} : \langle mq \rangle$. Because $mq \notin I^{k+1}$, $I^{k+1} : \langle mq \rangle \subsetneq \langle 1 \rangle$, i.e., $I^{k+1} : \langle mq \rangle$ is a proper monomial ideal of R, and in particular, it must be a subset of P. So $P = I^{k+1} : \langle mq \rangle$. But since $mq \in R \setminus I^{k+1}$, this implies that $P \in \mathrm{ass}(I^{k+1})$.

Remark 2.21 Herzog-Qureshi [107] called an ideal I *Ratliff* if $I^{k+1} : I = I^k$ for all $k \geq 1$. They show that if I is any ideal (not just a monomial ideal) that is Ratliff, then I has the persistence property.

Example 2.22 Theorem 2.20 does not classify ideals with the persistence property. As an example, consider the Stanley-Reisner ideal of the triangulation of the projective plane, i.e.,

$$I = \langle x_1x_2x_5, x_1x_3x_4, x_1x_2x_6, x_1x_3x_6, x_1x_4x_5,$$

$$x_2x_3x_4, x_2x_3x_5, x_2x_4x_6, x_3x_5x_6, x_4x_5x_6 \rangle.$$

Then a computer algebra program can show that $I^2 : I = I$, but $I^3 : I \neq I^2$, so I is not Ratliff. However, I has the persistence property.

While the above example shows that not every monomial ideal will satisfy $I^{k+1} : I = I^k$ for all k, Martínez-Bernal, Morey, and Villarreal [135] showed that equality holds for all edge ideals. Consequently, all edge ideals have the persistence property by Theorem 2.20. We sketch out the main ideas of this result since it is a nice example of using graph theory to prove a result in commutative algebra.

We first introduce the relevant graph theory.

Definition 2.23 A *matching* of a graph $G = (V, E)$ is a subset $\{e_1, \ldots, e_s\} \subseteq E$ such that $e_i \cap e_j = \emptyset$ for all $i \neq j$. The *matching number* is the size of the maximum matching of G; it is denoted $\alpha'(G)$. Note that if $|V| = n$, then $\alpha'(G) \leq \frac{n}{2}$. The *deficiency* of G, denoted $\mathrm{def}(G)$, is equal to $n - 2\alpha'(G)$.

Observe that $\mathrm{def}(G)$ counts the number of vertices in G that are not covered by any edge in a maximum matching.

We now introduce the operation of duplication.

Definition 2.24 If $G = (V, E)$ is a finite simple graph and $x \in V$, then the *duplication* of x in a graph G is the new graph $G' = (V', E')$ where $V' = V \cup \{x'\}$ for some new vertex x' and $E' = E \cup \{\{x', y\} \mid \{x, y\} \in E\}$.

In other words, if we duplicate a vertex x in a graph G, we are adding a new vertex x' and joining this new vertex to all the vertices to which x is adjacent. More

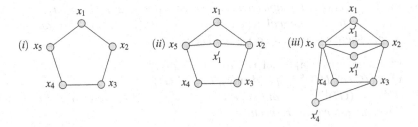

Fig. 2.3 Duplication: (i) Graph G, (ii) G with x_1 duplicated, and (iii) $G^{(3,1,1,2,1)}$

generally, if $\mathbf{a} = (a_1, \ldots, a_n) \in \mathbb{N}^n$, then $G^{\mathbf{a}}$ will denote the graph we obtain by duplicating vertex x_i successively $a_i - 1$ times. (If $a_i = 0$, then this means we delete the vertex x_i.)

Example 2.25 We illustrate the notation of duplication. Our initial graph G is the graph on the left in Fig. 2.3. The second graph (ii) shows the graph we obtain if we form the duplication of x_1. The final graph in (iii) shows the graph $G^{(3,1,1,2,1)}$. In this case, we duplicate the vertex x_1 twice, and we duplicate the vertex x_4 once, and leave the remaining vertices alone.

Give an edge $f = \{x_i, x_j\} \in E$, we shall let G^f denote the duplication of the vertices x_i and x_j. Note that if $\mathbf{e}_k$ is the standard basis vector of $\mathbb{N}^n$, then $G^f = G^{1+\mathbf{e}_i+\mathbf{e}_k}$ where $\mathbf{1}$ is the vector of all ones. With this notation, we can now state the required combinatorial result.

Theorem 2.26 ([135, Theorem 2.8]) *Let G be a graph. Then $\mathrm{def}(G^f) = \delta$ for all $f \in E$ if and only if $\mathrm{def}(G) = \delta$ and $\alpha'(G^f) = \alpha'(G) + 1$ for all $f \in E$.*

As an aside, to prove this combinatorial result, Martínez-Bernal et al. required a classical result on matchings attributed to Berge (see [135, Theorem 2.7]). For the interested reader, we have included this graph theory result.

Theorem 2.27 *For any graph G,*

$$\mathrm{def}(G) = \max\{c_0(G \setminus S) - |S| \mid S \subseteq V\}$$

where $c_0(-)$ denotes the number of connected components with an odd number of vertices.

The main insight of Martínez-Bernal et al. is that monomials in $I(G)^k \setminus I(G)^{k+1}$ are related to duplications of the graph G and the matching numbers of these new graphs. We record here only the statement of the technical lemma. Given a vector $\mathbf{a} = (a_1, \ldots, a_n) \in \mathbb{N}^n$ we use the notation $x^{\mathbf{a}}$ for $x_1^{a_1} \cdots x_n^{a_n}$, and $|\mathbf{a}| = a_1 + \cdots + a_n$.

Lemma 2.28 *Let G be a graph on n vertices with edge ideal $I(G) = \langle m_1, \ldots, m_t \rangle$ where the $m_i's$ are the minimal generators. Let $\mathbf{a} = (a_1, \ldots, a_n) \in \mathbb{N}^n$, and $m^{\mathbf{c}} = m_1^{c_1} \cdots m_t^{c_t}$ where $\mathbf{c} = (c_1, \ldots, c_t) \in \mathbb{N}^t$. Then*

(1) $x^{\mathbf{a}} = x^{\delta} m^{\mathbf{c}}$ where $|\delta| = \mathrm{def}(G^{\mathbf{a}})$ and $|\mathbf{c}| = \alpha'(G^{\mathbf{a}})$.

(2) $x^{\mathbf{a}} \in I(G)^k \setminus I(G)^{k+1}$ if and only if $k = \alpha'(G^{\mathbf{a}})$.

(3) $(G^{\mathbf{a}})^f = (G^{\mathbf{a}})^{\{x_i, x_j\}}$ for any edge $f = \{x_i', x_j''\}$ of $G^{\mathbf{a}}$ where x_i', respectively, x_j'' is a duplication of x_i, respectively, x_j.

We use the above results to now prove that all edge ideals have the persistence property.

Theorem 2.29 ([135, Corollary 2.17]) *For any graph G, the edge ideal $I(G)$ satisfies $I(G)^{k+1} : I(G) = I(G)^k$ for all $k \geq 1$. In particular, $I(G)$ has the persistence property.*

Proof If we can show that $I(G)^{k+1} : I(G) = I(G)^k$ for all $k \geq 1$, then $I(G)$ has the persistence property by Theorem 2.20. Since the containment $I(G)^k \subseteq I(G)^{k+1} : I(G)$ is straightforward to check, it suffices to show that $I(G)^{k+1} : I(G) \subseteq I(G)^k$.

Since $I(G)$ and $I(G)^{k+1}$ are monomial ideals, so is the colon ideal $I(G)^{k+1} : I(G)$. Let $\{m_1, \ldots, m_t\}$ be the minimal generators of $I(G)$, and consider any monomial $x^{\mathbf{a}} \in I(G)^{k+1} : I(G)$. Thus, $x^{\mathbf{a}} m_l \in I(G)^{k+1}$ for all $l = 1, \ldots, t$. Note that if $x^{\mathbf{a}} m_l \in I(G)^{k+2}$, then $x^{\mathbf{a}} \in I(G)^k$. So, we can assume that $x^{\mathbf{a}} m_l \in I(G)^{k+1} \setminus I(G)^{k+2}$. Because $m_l = x_i x_j$ for some i and j, we have $x^{\mathbf{a}} m_l = x^{\mathbf{a} + \mathbf{e}_i + \mathbf{e}_j}$. Thus by Lemma 2.28 (2), $\alpha'(G^{\mathbf{a} + \mathbf{e}_i + \mathbf{e}_j}) = k + 1$. Furthermore, since this is true for each m_i, this means that for each edge $f = \{x_i, x_j\}$ of G, $\alpha'((G^{\mathbf{a}})^{\{x_i, x_j\}}) = k + 1$.

Now for any edge $\{x_i', x_j''\}$ of $G^{\mathbf{a}}$ where x_i' and x_j'' are the duplicated vertices of x_i and x_j, Lemma 2.28 (3) gives

$$(G^{\mathbf{a}})^{\{x_i', x_j''\}} = (G^{\mathbf{a}})^{\{x_i, x_j\}}.$$

As a result, for all edges $f \in G^{\mathbf{a}}$ we have $\alpha'((G^{\mathbf{a}})^f) = k + 1$, and

$$\mathrm{def}((G^{\mathbf{a}})^f) = (|\mathbf{a}| + 2) - 2(k + 1) = |\mathbf{a}| - 2k.$$

By Theorem 2.26, we can deduce that $\mathrm{def}(G^{\mathbf{a}}) = |\mathbf{a}| - 2k$. By Lemma 2.28 (1), the monomial $x^{\mathbf{a}}$ is equal to $x^{\delta} m^{\mathbf{c}}$ where $|\delta| = \mathrm{def}(G^{\mathbf{a}})$ and $|\mathbf{c}| = \alpha'(G^{\mathbf{a}})$. Here $m^{\mathbf{c}} = m_1^{c_1} \cdots m_t^{c_t}$. Comparing the degrees of both sides of $x^{\mathbf{a}} = x^{\delta} m^{\mathbf{c}}$, we have

$$|\mathbf{a}| = (|\mathbf{a}| - 2k) + 2|\mathbf{c}| \Leftrightarrow k = |\mathbf{c}|.$$

But this means that $m^{\mathbf{c}} \in I(G)^k$, and thus $x^{\mathbf{a}} \in I(G)^k$, as desired.

Fig. 2.4 A graph whose
cover ideals fails the
persistence property;
originally discovered by
[123]

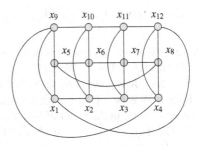

Theorem 2.29 shows that *all* edge ideals of graphs have the persistence property. This leads to the natural question of whether cover ideals have this property. For many large classes of graphs, this is indeed the case as shown by Francisco, Hà, and Van Tuyl.

Theorem 2.30 ([76]) *If G is a perfect graph, then J(G) has the persistence property.*

In fact, there are a number of graphs G that are not perfect, but $J(G)$ has the persistence property (e.g., the cover ideals of cycles). For a while, it was thought that cover ideals of all graphs (and in fact, all squarefree monomial ideals) satisfied the persistence property. Francisco, Hà, and Van Tuyl formulated a graph theory conjecture, that if true, would have implied the persistence property (see [74]). Interestingly, T. Kaiser, M. Stehlík, R. Škrekovski [123], all graph theorists, disproved this graph theory conjecture. The example, which is given below, is another nice illustration of the intersection between graph theory and commutative algebra.

Example 2.31 ([123]) The cover ideal of the graph G in Fig. 2.4 fails to have the persistence property. In particular, the maximal ideal is an associated prime of $J(G)^3$, but it is not an associated prime of $J(G)^4$. This graph G can be extended to an infinite family of graphs that fail to have the persistence property. The details are worked out in a paper of Hà and Sun [93].

One obvious open question is the following:

Question 2.32 Classify all finite simple graphs G whose cover ideal fails to have the persistence property.

It should be noted that there are some other families of squarefree monomial ideals (that are neither edge ideals or cover ideals) that are known to satisfy the persistence property. This includes polymatroidal ideals studied by Herzog, Rauf, and Vladoiu [111] and some generalized cover ideals studies by Bhat, Biermann, and Van Tuyl [16].

2.5 Elements of ass(I^s)

We now turn to the problem of determining the elements of ass(I^s), i.e., the focus of Question 2.3.(iii). We will focus on the cover ideals of graphs, although many of these ideas extend to all squarefree monomial ideals using the language of hypergraphs (see [76] for more details). Some of the material of this section can also be found in the paper of Van Tuyl [164].

The key idea that you should take away is that if $P = \langle x_{i_1}, \ldots, x_{i_r} \rangle \in$ ass($J(G)^s$), then something "interesting" is happening on the induced graph G_P, where we view P as both an ideal generated by the variables and as a subset $P \subseteq V(G)$. Here G_P is the induced graph on P, that is, the graph with the vertex set P and edge set $E(G_P) = \{e \in E(G) \mid e \subseteq P\}$. Specifically, associated primes are related to colourings of the graph.

The following theorem of Francisco, Hà, and Van Tuyl [76], which is interesting in its own right, shows that the chromatic number is related to powers of ideals.

Theorem 2.33 ([76]) *For any graph G on n vertices,*

$$\chi(G) = \min\{d \mid (x_1 \cdots x_n)^{d-1} \in J(G)^d\}.$$

Proof For any subset $W \subseteq V(G)$, we set $x^W = \prod_{x_i \in W} x_i$ in $R = \mathbb{K}[x_1, \ldots, x_n]$.

($\Rightarrow$) Suppose that $(x_1 \cdots x_n)^{d-1} \in J(G)^d$. Then there exists d minimal vertex covers $W_1, \ldots, W_d$ (not necessarily distinct) such that $x^{W_1} \cdots x^{W_d} \mid (x_1 \cdots x_n)^{d-1}$. For each $x_i \in \{x_1, \ldots, x_n\}$, there exists some W_j such that $x_i \notin W_j$; otherwise, if $x_i \in W_j$ for all $1 \le j \le d$, then the power of x_i is d in $x^{W_1} \cdots x^{W_d}$, from which it follows that $x^{W_1} \cdots x^{W_d}$ cannot divide $(x_1 \cdots x_n)^{d-1}$, a contradiction.

Now form the following d sets:

$$C_1 = V \setminus W_1$$
$$C_2 = (V \setminus W_2) \setminus C_1$$
$$C_3 = (V \setminus W_3) \setminus (C_1 \cup C_2)$$
$$\vdots$$
$$C_d = (V \setminus W_d) \setminus (C_1 \cup \cdots \cup C_{d-1}).$$

It suffices to show that $C_1, \ldots, C_d$ form a d-colouring of G. We first note that by construction, the C_is are pairwise disjoint. As well, because each $C_i \subseteq V \setminus W_i$, each C_i is an independent set. So it remains to show that $V = C_1 \cup \cdots \cup C_d$. If $x \in V$, there exists some W_j such $x \notin W_j$, whence $x \in V \setminus W_j$. Hence $x \in C_j$ or $x \in (C_1 \cup \cdots \cup C_{j-1})$. Thus $\chi(G) \le d$.

($\Leftarrow$) It suffices to show that $(x_1 \cdots x_n)^{\chi(G)-1} \in J(G)^{\chi(G)}$. Let $C_1 \cup \cdots \cup C_{\chi(G)}$ be a $\chi(G)$-colouring of G. For each $i = 1, \ldots, \chi(G)$, set

$$Y_i = C_1 \cup \cdots \cup \widehat{C_i} \cup \cdots \cup C_{\chi(G)}.$$

Since $Y_i = V \setminus C_i$, and because C_i is an independent set, the set Y_i is a vertex cover of G. Hence $x^{Y_i} \in J(G)$ for $i = 1, \ldots, \chi(G)$. It follows that

$$\prod_{i=1}^{\chi(G)} x^{Y_i} = \left(\prod_{i=1}^{\chi(G)} x^{C_i} \right)^{\chi(G)-1} = (x_1 \cdots x_n)^{\chi(G)-1} \in J(G)^{\chi(G)}.$$

The next lemma is similar to the above result, and will be used below.

Lemma 2.34 *Let G be a finite simple graph on $V(G) = \{x_1, \ldots, x_n\}$ with cover ideal $J(G)$. Suppose that for some independent set $C \subseteq V$, the monomial $(x_1 \cdots x_n)^{d-1} x^C \in J(G)^d$. Then $\chi(G) \leq d + 1$.*

Proof Let $W = V \setminus C$. Since C is an independent set, W is a vertex cover of G. It follows that $(x_1 \cdots x_n)^d = ((x_1 \cdots x_n)^{d-1} x^C) x^W \in J(G)^{d+1}$. By Theorem 2.33, this implies that $\chi(G) \leq d + 1$.

We need to recall some more graph theory. Below, if G is a graph with $x \in V(G)$, then $G \setminus \{x\}$ denotes the graph one obtains by removing x and all edges containing x from G.

Definition 2.35 A graph G is *critically s-chromatic* if $\chi(G) = s$, and for every $x \in V(G)$, $\chi(G \setminus \{x\}) < s$.

Example 2.36 Let $G = C_n$ be the n-cycle with n odd. Then G is a critically 3-chromatic graph since $\chi(G) = 3$, but if we remove any vertex x, $\chi(G \setminus \{x\}) = 2$.

Example 2.37 Let $G = K_n$ be the clique of size n. Then G is a critically n-chromatic graph since $\chi(G) = n$, but if we remove any vertex x, $G \setminus \{x\} = K_{n-1}$, and thus $\chi(G \setminus \{x\}) = n - 1$.

Remark 2.38 In the above examples, the chromatic number of $G \setminus \{x\}$ is one less than $\chi(G)$ for each vertex x. This holds in general, i.e., if G is critically s-chromatic, then $\chi(G \setminus \{x\}) = s - 1$ for all $x \in V(G)$. Note that the definition implies that $\chi(G \setminus \{x\}) \leq s - 1$. Suppose that $\chi(G \setminus \{x\}) < s - 1$. The colouring of $G \setminus \{x\}$ combined with a distinct colour given to x gives an $s - 1$ colouring of G, contradicting the fact that $\chi(G) = s$. So $\chi(G \setminus \{x\}) = s - 1$.

Remark 2.39 You should be able to convince yourself that the only critically 1-chromatic graph is the graph of an isolated vertex, and the only critically 2-chromatic graph is K_2. The only critically 3-chromatic graphs are precisely the graphs $G = C_n$ with n odd. However, for $s \geq 4$, there is no known classification of critically s-chromatic graphs.

As we saw in Lemma 2.4, when I is a monomial ideal and $P \in \text{ass}(I)$, then there is a monomial m such that $I : \langle m \rangle = P$. When I is the power of a cover ideal, we can deduce some further information about the monomial m.

Lemma 2.40 *Let G be a finite simple graph on $V = \{x_1, \ldots, x_n\}$ with cover ideal $J(G)$. Suppose that $\langle x_1, \ldots, x_n \rangle \in \text{ass}(J(G)^d)$. If m is such that $J(G)^d : \langle m \rangle = \langle x_1, \ldots, x_n \rangle$, then $m \mid (x_1 \cdots x_n)^{d-1}$.*

Proof If $m \nmid (x_1 \cdots x_n)^{d-1}$, then there exists some x_i in m whose exponent is at least d. Thus, the exponent of x_i in mx_i is at least $d + 1$. Because $mx_i \in J(G)^d$, there exist generators $m_1, \ldots, m_d \in J(G)$ (not necessarily distinct) such that

$$mx_i = m_1 \cdots m_d N$$

for some monomial N. Because each m_i is squarefree, the exponent of x_i in m_i is at most one. So $x_i \mid N$, whence $m = m_1 \cdots m_d(\frac{N}{x_i}) \in J(G)^d$, a contradiction since $m \notin J(G)^d$.

As the next theorem shows, some of the associated primes of $J(G)^s$ are actually detecting induced subgraphs that are critically $(s + 1)$-chromatic.

Theorem 2.41 *Let G be a graph and suppose $P \subseteq V(G)$ is such that G_P is critically $(s + 1)$-chromatic. Then*

(1) $P \notin \text{ass}(J(G)^d)$ *for* $1 \leq d < s$.
(2) $P \in \text{ass}(J(G)^s)$.

Proof By Lemma 2.8, we can assume that $G = G_P$ and $P = \langle x_1, \ldots, x_n \rangle$.

(1) Suppose that $P = \langle x_1, \ldots, x_n \rangle \in \text{ass}(J(G)^d)$ with $1 \leq d < s$. Then there exists some monomial m such that $J(G)^d : \langle m \rangle = \langle x_1, \ldots, x_n \rangle$. By Lemma 2.40, $m \mid (x_1 \cdots x_n)^{d-1}$, i.e., the exponent of each x_i in m is at most $d - 1$. On the other hand, if $mx_i \mid (x_1 \cdots x_n)^{d-1}$ for some i, then since $mx_i \in J(G)^d$, we would have $(x_1 \cdots x_n)^{d-1} \in J(G)^d$, whence $\chi(G) \leq d$ by Theorem 2.33, a contradiction. Thus, $mx_i \nmid (x_1 \cdots x_n)^{d-1}$, and hence, $m = (x_1 \cdots x_n)^{d-1}$. So $mx_i = (x_1 \cdots x_n)^{d-1}x_i \in J(G)^d$ for each $i = 1, \ldots, n$. We now apply Lemma 2.34 with $C = \{x_i\}$ to conclude that $\chi(G) \leq d + 1 < s + 1 = \chi(G)$. But this is a contradiction. Hence, $P \notin \text{ass}(J^d)$ if $1 \leq d < s$.
(2) We are given

$$\chi(G) = \min\{d \mid (x_1 \cdots x_n)^{d-1} \in J(G)^d\} = s + 1$$

so $m = (x_1 \cdots x_n)^{s-1} \notin J(G)^s$. In other words, we have, $J(G)^s : \langle m \rangle \subsetneq \langle 1 \rangle$, and hence $J(G)^s : \langle m \rangle \subseteq \langle x_1, \ldots, x_n \rangle$. We will now show that $J(G)^s : \langle m \rangle \supseteq \langle x_1, \ldots, x_n \rangle$; the conclusion will then follow from this fact.

Since $\chi(G)$ is critically $(s+1)$-chromatic, by Remark 2.38 we have $\chi(G\setminus\{x_i\}) = s$ for each $x_i \in V(G)$. Let

$$V(G \setminus \{x_i\}) = C_1 \cup \cdots \cup C_s$$

be the s colouring of $V(G \setminus \{x_i\})$ where C_i denotes all the vertices coloured i. Then

$$V(G) = C_1 \cup \cdots \cup C_s \cup \{x_i\}$$

is an $(s + 1)$-colouring of G.

For $j = 1, \ldots, s$, set

$$W_j = C_1 \cup \cdots \cup \widehat{C_j} \cup \cdots \cup C_s \cup \{x_i\}.$$

Each W_j is a vertex cover, so $x^{W_j} \in J(G)$. Thus

$$\prod_{j=1}^{s} x^{W_j} \in J(G)^s.$$

But $\prod_{j=1}^{s} x^{W_j} = (x_1 \cdots x_n)^{s-1} x_i$. Thus, $x_i \in J(G)^s : \langle m \rangle$. This is true for each $x_i \in V(G)$, whence $\langle x_1, \ldots, x_n \rangle \subseteq J(G)^s : \langle m \rangle \subseteq \langle x_1, \ldots, x_n \rangle$, as desired.

Example 2.42 We consider the following graph

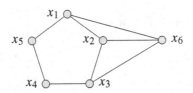

Note that the induced graph on $\{x_1, x_2, x_6\}$ is a K_3 (and C_3), a critically 3-chromatic graph. So $P = \langle x_1, x_2, x_6 \rangle$ is in $\mathrm{Ass}(J(G)^2)$, but not in $\mathrm{Ass}(J(G))$. Similarly, since the induced graph on $\{x_1, x_2, x_3, x_4, x_5\}$ is a C_5, we will have $\langle x_1, x_2, x_3, x_4, x_5 \rangle \in \mathrm{Ass}(J(G)^2)$.

When $s = 2$, we can find a converse of Theorem 2.41. A complete characterization of the associated primes of $J(G)^2$ was first given in the Francisco, Hà, and Van Tuyl [75].

Theorem 2.43 *For any graph G, $P \in \mathrm{ass}(J(G)^2)$ if and only if*

(i) $P = \langle x_i, x_j \rangle$ and $\{x_i, x_j\} \in E(G)$, or
(ii) $P = \langle x_{i_1}, \ldots, x_{i_r} \rangle$ where r is odd and $G_P = C_r$, an odd cycle.

Unfortunately, the converse of Theorem 2.41 is false in general; that is, if $P \in \mathrm{ass}(J(G)^s)$, but $P \notin \mathrm{ass}(J(G)^d)$ with $1 \le d < s$, then the graph G_P is not necessarily a critically $(s+1)$-chromatic graph.

Example 2.44 Consider the graph of Example 2.42. Then the prime ideal $P = \langle x_1, x_2, x_3, x_4, x_5, x_6 \rangle \in \mathrm{ass}(J(G)^3)$, but not in $\mathrm{ass}(J(G))$ or $\mathrm{ass}(J(G)^2)$. However, the graph $G = G_P$ is not critically 4-chromatic. In fact, $\chi(G) = 3$.

What is happening here is that we are looking in the "wrong" graph. We need to consider a larger graph that contains our initial graph as a subgraph. This approach is similar to the approach we introduced in the previous section when we described the duplication of a vertex to study powers of $I(G)$. To study powers of $J(G)$, we need a similar construction.

Definition 2.45 Given a graph $G = (V(G), E(G))$ and integer $s \ge 1$, the *s-th expansion of G*, denoted G^s, is the graph constructed from G as follows: (a) replace each $x_i \in V(G)$ with a clique of size s on the vertices $\{x_{i,1}, \ldots, x_{i,s}\}$, and (b) two vertices $x_{i,a}$ and $x_{j,b}$ are adjacent in G^s if and only if x_i and x_j were adjacent in G.

Example 2.46 We illustrate this example when $G = C_4$, and we construct G^2. Recall that C_4 is the graph:

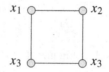

Then the second expansion of G is the graph:

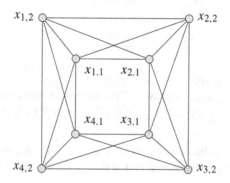

The following result of Francisco, Hà, and Van Tuyl gives a combinatorial interpretation for the elements of $\mathrm{ass}(J(G)^s)$ in terms of colourings and expansions of the graph G.

Theorem 2.47 ([76])   *Let G be a graph with cover ideal $J(G)$. Then $\langle x_{i_1}, \ldots, x_{i_r} \rangle \in \mathrm{ass}(J(G)^s)$ if and only if there exists a set $T \subseteq V(G^s)$ with*

$$\{x_{i_1,1}, x_{i_2,1}, \ldots, x_{i_r,1}\} \subseteq T \subseteq \{x_{i_1,1}, \ldots, x_{i_1,s}, \ldots, x_{i_r,1}, \ldots, x_{i_r,s}\}$$

such that the induced graph $(G^s)_T$ is a critically $(s+1)$-chromatic graph.

The proof is a mixture of a number of ingredients. First, instead of looking at the primary decomposition, one considers the irreducible decomposition of $J(G)^s$. Then one uses tools such as generalized Alexander duality, polarization and depolarization of monomial ideals, and a result of Sturmfels and Sullivant [156]. We have only stated the result for cover ideals of graphs, but the result holds also for cover ideals of hypergraphs, i.e., any squarefree monomial ideal.

Example 2.48 Let us return to Example 2.42 and explain why $\langle x_1, x_2, x_3, x_4, x_5, x_6 \rangle$ appears in $\mathrm{ass}(J(G)^3)$. We form G^3, the 3-rd expansion of G. If we consider the induced subgraph on $T = \{x_{1,1}, x_{2,1}, x_{3,1}, x_{4,1}, x_{5,1}, x_{5,2}, x_{6,1}\} \subseteq V(G^3)$, we find the graph

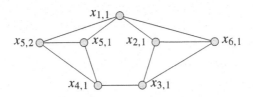

You can now convince yourself that this graph is critically 4-chromatic. Consequently, $\langle x_1, x_2, x_3, x_4, x_5, x_6 \rangle \in \mathrm{ass}(J(G)^3)$.

Remark 2.49 The work of Francisco, Hà, and Van Tuyl on the associated primes of I^s when I is a squarefree monomial ideal takes the point of view that I was the cover ideal of a hypergraph, and consequently, the generators correspond to minimal vertex covers. Hien, Lam, and Trung [112] took an alternative point-of-view. They viewed the generators of I as the edges of a (hyper)graph, and described the associated primes in terms of this (hyper)graph.

Remark 2.50 Bayati, Herzog, and Rinaldo [12] have shown that for any monomial ideal I, there is an algorithm to find all the primes in $\mathrm{ass}(I^{\mathrm{astab}(I)})$ using Koszul homology.

As we have shown, there has been progress on the problems raised in Question 2.3. However, there are still a number of interesting problems that remain open (including some of the problems distributed at PRAGMATIC; see Chap. 19). We leave this chapter with another question that is based upon computer experiments of Francisco, Hà, and Van Tuyl. Specifically:

Question 2.51 Let I be any squarefree monomial ideal. If $P \in \mathrm{ass}(I^2)$, is $P \in \mathrm{ass}(I^s)$ for all $s \geq 2$?

Note that when I is the cover ideal of a graph, then it is known that $\text{ass}(I^2) \subseteq \text{ass}(I^n)$ for all $n \geq 2$. Furthermore, by the work of Martínez-Bernal, Morey, and Villarreal, we also know it is true for all edge ideals. The above question asks if this behaviour is true for all squarefree monomial ideals.

Chapter 3
Final Comments and Further Reading

As we have hopefully demonstrated in the last two chapters, Question 1.2 has motivated a number of interesting results, including some nice connections with combinatorics. Although we cannot cover all of the existing literature, here are some suggested references for further reading.

The monograph of McAdams [136] gives a summary of many of the results that were discovered in the 1970s by people such as Brodmann, Ratliff, and McAdams. Ratliff [148] provides a historical introduction to the topic. Note that in older papers, Question 1.2 is sometimes expressed as understanding the asymptotic prime divisors of an ideal. Swanson's lecture notes [159] also provide an introduction to these results. It works out many of the details that were explained in the first chapter.

In the past decade, Question 1.2 has been revisited primarily in the case that I is a monomial ideal. As shown in the previous chapter, when I is a monomial ideal, one may be able to associate to I a combinatorial object that may encode (or relates to) the elements of ass(I^s). One of the first papers taking this approach is a paper of Chen, Morey, and Sun [37] which considers the case that $I = I(G)$ is an edge ideal. Among other things, they show that the existence of certain structures in the graph G forces the existence of elements in ass($I(G)^s$). Martínez-Bernal, Morey, and Villarreal [135] later showed that all edge ideals $I(G)$ have the persistence property. Recently, Lam and Trung [128] have shown how to describe all the elements of ass($I(G)^s$) in terms of a graph theoretical tool called the ear decomposition of a graph. This work builds upon earlier work of Hien, Lam, and Trung [112] and Hien and Lam [112]. Terai and Trung [161] and Francisco, Hà, and Van Tuyl [75] both consider the special case of ass(I^2) when I is an ideal constructed from a graph or a hypergraph.

There is a dual notion of an edge ideal, called the cover ideal. Taking the point-of-view that a monomial ideal I is a cover ideal of a hypergraph, Francisco, Hà, and Van Tuyl [76] described all other elements of ass(I^s) in terms of colourings of the associated hypergraph (and hypergraphs built from the original hypergraph). This approach was described in this chapter; also see the survey article of Francisco,

E. Carlini et al., *Ideals of Powers and Powers of Ideals*, Lecture Notes of the Unione Matematica Italiana 27, https://doi.org/10.1007/978-3-030-45247-6_3

Hà, and Mermin [77]. The cover ideal point-of-view was also used by Bhat, Biermann, and Van Tuyl [16] to describe a family of ideals constructed from trees with the persistence property and compute its index of stability. Nasernejad and Khashyarmanesh [142] considered similar ideals to find another family of ideals with the persistence property.

One can also consider Question 1.2 for arbitrary monomial ideals, not just squarefree monomial ideals. The papers of Herzog and Qureshi [107] and Herzog, Rauf, and Vladoiu [111] identify some families of monomial ideals (most notably, the family of polymatroidal ideals) that have the persistence property. In addition, they also have some results on the index of stability for these families and the stable set of the associated primes (the set $\mathrm{ass}(I^{s_0})$ of Theorem 1.4). Khashyarmanesh and Nasernejad [126] have also looked at the stable set $\mathrm{ass}(I^{s_0})$; in particular, they show that for any set of prime monomial ideals $\mathscr{P} = \{P_1, \ldots, P_s\}$ in $\mathbb{K}[x_1, \ldots, x_n]$, there is a monomial ideal I with $\mathrm{ass}(I^{s_0}) = \mathscr{P}$. When I is a squarefree monomial ideal, Bayati, Herzog, and Rinaldo [12] have shown how to compute $\mathrm{ass}(I^{s_0})$ (complete with computer code).

One of the PRAGMATIC projects looked for new examples of monomial ideals that failed to have the persistence property. In particular, Bela, Favacchio, and Tran [13] leveraged the examples of Kaiser, Stehlík, and Škrekovski [123] to build new monomial ideals corresponding to hypergraphs that failed to have this property.

Part II
Regularity of Powers of Ideals

Chapter 4
Regularity of Powers of Ideals and the Combinatorial Framework

Castelnuovo-Mumford regularity (or simply *regularity*) is an important invariant in commutative algebra and algebraic geometry. Computing or finding bounds for the regularity is a difficult problem. In the next three chapters, we shall address the regularity of ordinary and symbolic powers of squarefree monomial ideals.

Our interest in squarefree monomial ideals comes from their strong connections to topology and combinatorics via the construction of Stanley-Reisner ideals and edge ideals. In recent years advances in computer technology and speed of computation have drawn significant attention toward problems and questions involving this class of ideals.

The collection of problems and questions presented in these three chapters originates from a celebrated result proved independently by Cutkosky, Herzog and Trung [48] and Kodiyalam [127] (see also Trung and Wang [162] for the module case and Bagheri, Chardin, and Hà,[6] and Whieldon [170] for the multigraded case), which states that for a homogeneous ideal I in a standard graded algebra R over a Noetherian commutative ring, the regularity of I^q is asymptotically a linear function. The problem of determining this linear function and the smallest value of q starting from which reg I^q becomes linear remains wide open and has evolved into a highly active research area in the last few decades.

We shall discuss this problem primarily for the class of squarefree monomial ideals. Our focus will be on studies of the asymptotic linear function reg I^q for a squarefree monomial ideal I via combinatorial data and structures of the corresponding simplicial complex and/or hypergraph.

4.1 Regularity of Powers of Ideals: The General Question

The main object of our discussion in this part of the book is the Castelnuovo-Mumford regularity. This notion can be defined in various ways. We shall first give the definition for modules over polynomial rings as this situation is our focus. A

© The Editor(s) (if applicable) and The Author(s), under exclusive licence to Springer Nature Switzerland AG 2020
E. Carlini et al., *Ideals of Powers and Powers of Ideals*, Lecture Notes of the Unione Matematica Italiana 27, https://doi.org/10.1007/978-3-030-45247-6_4

more general definition in terms of local cohomology will also be given for the more advanced interested reader. The motivating theorem and general question are given at the end of the section.

Definition 4.1 Let R be a standard graded polynomial ring over a field and let $\mathfrak{m}$ be its maximal homogeneous ideal. Let M be a finitely generated graded R-module and let

$$0 \to \bigoplus_{j \in \mathbb{Z}} R(-j)^{\beta_{p,j}(M)} \to \cdots \to \bigoplus_{j \in \mathbb{Z}} R(-j)^{\beta_{0,j}(M)} \to M \to 0$$

be its minimal free resolution. Then the regularity of M is given by

$$\operatorname{reg} M = \max\{j - i \mid \beta_{i,j}(M) \neq 0\}.$$

Remark 4.2 It is clear from the definition that the regularity of M gives an upper bound for the generating degrees of M.

Example 4.3 Consider

$$I = \langle x^2 y - 2yz^2 + 3z^3, 2xw - 3yw, yw^4 - y^2z^3 - 2x^5 \rangle \subseteq R = \mathbb{Q}[x, y, z, w].$$

Then I has the following minimal free resolution:

$$0 \longrightarrow R(-10) \longrightarrow R(-5) \oplus R(-7) \oplus R(-8) \longrightarrow R(-2) \oplus R(-3) \oplus R(-5) \longrightarrow I \longrightarrow 0.$$

Thus, $\operatorname{reg} I = 8$.

If R is a general standard graded algebra over a ring, then the minimal free resolution of an R-module M may not be finite. In this case, the regularity can still be defined via local cohomology. See, for example, Chardin [35], and Eisenbud and Goto [64] for the equivalence between the two definitions when R is a polynomial ring over a field.

Definition 4.4 Let R be a standard graded algebra over a Noetherian commutative ring with identity and let $\mathfrak{m}$ be its maximal homogeneous ideal. Let M be a finitely generated graded R-module. For $i \geq 0$, let

$$a^i(M) = \begin{cases} \max\left\{ l \in \mathbb{Z} \,\middle|\, \left[H^i_{\mathfrak{m}}(M)\right]_l \neq 0 \right\} & \text{if } H^i_{\mathfrak{m}}(M) \neq 0 \\ -\infty & \text{otherwise.} \end{cases}$$

The *regularity* of M is defined to be

$$\operatorname{reg} M = \max_{i \geq 0}\{a^i(M)\}.$$

Note that $a^i(M) = 0$ for $i > \dim M$, so the regularity of M is well-defined.

Example 4.5 Consider $R = \mathbb{K}[x_1, \ldots, x_n]$, a polynomial ring over a field $\mathbb{K}$. Then $H_\mathfrak{m}^i(R) = 0$ for all $i < n$, and $a^n(R) = -n$. Thus, reg $R = 0$.

This definition of regularity works especially well with short exact sequences. For instance, the following lemma is well-known (cf. [63, Corollary 20.19]).

Lemma 4.6 *Let* $0 \to M \to N \to P \to 0$ *be a short exact sequence of graded R-modules. Then*

1. reg $N \leq \max\{\text{reg } M, \text{reg } P\}$,
2. reg $M \leq \max\{\text{reg } N, \text{reg } P + 1\}$,
3. reg $P \leq \max\{\text{reg } M - 1, \text{reg } N\}$,
4. reg $M = \text{reg } P + 1$ *if* reg $N < \text{reg } P$,
5. reg $P = \text{reg } M - 1$ *if* reg $N < \text{reg } M$,
6. reg $P = \text{reg } N$ *if* reg $N > \text{reg } M$, *and*
7. reg $N = \text{reg } M$ *if* reg $P + 1 < \text{reg } M$.

The motivation of our discussion is the following celebrated result, which was first independently proved by Cutkosky, Trung and Herzog [48] and Kodiyalam [127] (the constant a was determined in Trung and Wang [162]).

Theorem 4.7 ([48, 127, 162]) *Let R be a standard graded algebra over a Noetherian commutative ring with identity. Let $I \subseteq R$ be a homogeneous ideal. Then there exist constants a and b such that*

$$\text{reg } I^q = aq + b \text{ for all } q \gg 0.$$

Moreover,

$$a = \min\{d(J) \mid J \text{ is a minimal homogeneous reduction of } I\}.$$

Here, $J \subseteq I$ is a *reduction* of I if $I^{s+1} = JI^s$ for some (and all) $s \geq 0$, and $d(J)$ denotes the maximal generating degree of J.

The following problem remains wide open despite much effort from researchers.

Problem 4.8 Determine b and $q_0 = \min\{t \in \mathbb{Z} \mid \text{reg } I^q = aq + b \text{ for all } q \geq t\}$.

In general, when I is generated in the same degree, the constant b can be related to a *local* invariant, namely, the regularity of preimages of germs of schemes via certain projection maps from the blowup of $X = \text{Proj } R$ along I.

For the interested reader, we expanded upon the above comment; we do not refer to this discussion in future sections. Let $I = \langle F_0, \ldots, F_m \rangle$, where $F_0, \ldots, F_m$ are homogeneous elements of degree d in R. Let $\pi : \tilde{X} \to X$ be the blow up of $X = \text{Proj } R \subseteq \mathbb{P}^n$ along the subscheme defined by I. Let $\mathscr{R} = R[It] = \bigoplus_{q \geq 0} I^q t^q$ be the Rees algebra of I. By letting $\deg t = (0, 1)$ and $\deg F_i t = (d, 1)$, the Rees algebra $\mathscr{R}$ is naturally bi-graded with $\mathscr{R} = \bigoplus_{p,q \in \mathbb{Z}} \mathscr{R}_{(p,q)}$, where $\mathscr{R}_{(p,q)} = (I^q)_{p+qd} t^q$. Under this bi-gradation of $\mathscr{R}$, we can define the bi-projective scheme

$\operatorname{Proj}\mathscr{R}$ of $\mathscr{R}$ as follows (cf. Hà [92]):

$$\operatorname{Proj}\mathscr{R} = \{\mathfrak{p} \in \operatorname{Spec}\mathscr{R} \mid \mathfrak{p} \text{ is a bihomogeneous ideal and } \mathscr{R}_{++} \not\subseteq \mathfrak{p}\},$$

where $\mathscr{R}_{++} = \bigoplus_{p,q\geq 1} \mathscr{R}_{(p,q)}$. It can be seen that $\operatorname{Proj}\mathscr{R} \subseteq \mathbb{P}^n \times \mathbb{P}^m$ and $\tilde{X} \simeq \operatorname{Proj}\mathscr{R}$.

Let $\phi : \operatorname{Proj}\mathscr{R} \to \mathbb{P}^m$ denote the natural projection from $\operatorname{Proj}\mathscr{R}$ onto its second coordinate, and let $\overline{X} = \operatorname{im}(\phi)$. Note that ϕ is the morphism given by the divisor $D = dE_0 - E$, where E is the exceptional divisor of π and E_0 is the pullback of a general hyperplane in X. For a point $\wp \in \overline{X}$, let $\tilde{X}_\wp = \tilde{X} \times_{\overline{X}} \operatorname{Spec}\mathscr{O}_{\overline{X},\wp}$ be the preimage of ϕ over the affine scheme $\operatorname{Spec}\mathscr{O}_{\overline{X},\wp}$.

Let S denote the homogeneous coordinate ring of $\overline{X} \subseteq \mathbb{P}^m$. For a homogeneous prime $\wp \subseteq S$ (i.e., a point in $\overline{X}$), let $\mathscr{R}_\wp = \mathscr{R} \otimes_S S_\wp$ be the *localization* of $\mathscr{R}$ at $\wp$. The *homogeneous localization* of $\mathscr{R}$ at $\wp$, denoted by $\mathscr{R}_{(\wp)}$, is defined to be the collection of elements in $\mathscr{R}_\wp$ of degree 0 in terms of powers of t. Then $\tilde{X}_\wp = \operatorname{Proj}\mathscr{R}_{(\wp)}$. We define the *regularity* of $\tilde{X}_\wp$, denoted by $\operatorname{reg}\tilde{X}_\wp$, to be that of its homogeneous coordinate ring $\mathscr{R}_{(\wp)}$, and let $\operatorname{reg}\phi = \max\{\operatorname{reg}\tilde{X}_\wp \mid \wp \in \overline{X}\}$.

The following result follows from a series of work of Chardin [36], Eisendbud and Harris [65], and Hà [92]. Partial results on the stability index q_0 were obtained by Eisenbud and Ulrich [66], when I is $\mathfrak{m}$-primary, and by Chardin [35] and Bisui, Hà, and Thomas [18], when I is equi-generated.

Theorem 4.9 *Let R be a standard graded algebra over a Noetherian commutative ring with identity. Let $I \subseteq R$ be a homogeneous ideal generated in degree d. For $q \gg 0$, we have*

$$\operatorname{reg} I^q = qd + \operatorname{reg}\phi.$$

The invariant $\operatorname{reg}\phi$, in practice, is difficult to compute. Even when I is generated by "enough" (i.e., more than $\dim R$) general linear forms, it is still an open problem to compute $\operatorname{reg}\phi$.

In the next three chapters, we shall see a different approach to computing $\operatorname{reg}\phi$ (or equivalently, the free constant b) when I is a squarefree monomial ideal.

4.2 Squarefree Monomial Ideals and Combinatorial Framework

Our aim in this part of the book is to study a restricted version of Problem 4.8, which is applied to the class of squarefree monomial ideals. For this purpose, we shall now fix some notation. From now on, $\mathbb{K}$ will denote an infinite field, $R = \mathbb{K}[x_1, \ldots, x_n]$ will be a polynomial ring over $\mathbb{K}$, and $\mathfrak{m}$ will denote the maximal homogeneous ideal in R. For obvious reasons, we shall identify the variables $x_1, \ldots, x_n$ with the vertices of simplicial complexes and hypergraphs being discussed. By abusing notation, we also often identify a subset V of the vertices $X = \{x_1, \ldots, x_n\}$ with the squarefree monomial $x^V = \prod_{x \in V} x$ in the polynomial ring R.

The combinatorial framework we shall use is the construction of Stanley-Reisner ideals and edge ideals corresponding to simplicial complexes and hypergraphs. The notion of edge ideals of hypergraphs is the generalization of that of edge ideals of graphs defined in Chap. 2.

4.2.1 Simplicial Complexes

A *simplicial complex* Δ over the vertex set $X = \{x_1, \ldots, x_n\}$ is a collection of subsets of X such that if $F \in \Delta$ and $G \subseteq F$, then $G \in \Delta$. Elements of Δ are called *faces*. Maximal faces (with respect to inclusion) are called *facets*. For $F \in \Delta$, the *dimension* of F is defined to be $\dim F = |F| - 1$. The *dimension* of Δ is $\dim \Delta = \max\{\dim F \mid F \in \Delta\}$. The complex is called *pure* if all of its facets are of the same dimension. A graph can be viewed as a 1-dimensional simplicial complex.

Let Δ be a simplicial complex, and let $Y \subseteq X$ be a subset of its vertices. The *induced subcomplex* of Δ on Y, denoted by $\Delta[Y]$, is the simplicial complex with vertex set Y and faces $\{F \in \Delta \mid F \subseteq Y\}$.

Definition 4.10 Let Δ be a simplicial complex over the vertex set X, and let $\sigma \in \Delta$.

1. The *deletion* of σ in Δ, denoted by $\mathrm{del}_\Delta(\sigma)$, is the simplicial complex obtained by removing σ and all faces containing σ from Δ.
2. The *link* of σ in Δ, denoted by $\mathrm{link}_\Delta(\sigma)$, is the simplicial complex whose faces are

$$\{F \in \Delta \mid F \cap \sigma = \emptyset, \sigma \cup F \in \Delta\}.$$

Definition 4.11 A simplicial complex Δ is recursively defined to be *vertex decomposable* if either

1. Δ is a simplex (or the empty simplicial complex); or
2. there is a vertex v in Δ such that both $\mathrm{link}_\Delta(v)$ and $\mathrm{del}_\Delta(v)$ are vertex decomposable, and all facets of $\mathrm{del}_\Delta(v)$ are facets of Δ.

A vertex satisfying condition (2) is called a *shedding vertex*, and the recursive choice of shedding vertices are called a *shedding order* of Δ.

Definition 4.12 A simplicial complex Δ is said to be *shellable* if there exists a linear order of its facets $F_1, F_2, \ldots, F_t$ such that for all $k = 2, \ldots, t$, the subcomplex $\left(\bigcup_{i=1}^{k-1} \overline{F_i}\right) \cap \overline{F_k}$ is pure and of dimension $(\dim F_k - 1)$. Here $\overline{F}$ represents the simplex over the vertices of F.

Fig. 4.1 A vertex
decomposable simplicial
complex

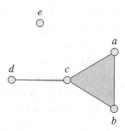

It is a celebrated fact that *pure* shellable complexes give rise to *Cohen-Macaulay Stanley-Reisner rings*. For more details on Cohen-Macaulay rings and modules, we refer the reader to Bruns and Herzog [25]. The notion of Stanley-Reisner rings will be discussed later in the section. Note also that a ring or module is *sequentially Cohen-Macaulay* if it has a filtration in which the factors are Cohen-Macaulay and their dimensions are increasing. This property corresponds to (*nonpure*) shellability in general.

Vertex decomposability can be thought of as a combinatorial criterion for shellability and sequentially Cohen-Macaulayness. In particular, for a simplicial complex Δ,

$$\Delta \text{ vertex decomposable} \Rightarrow \Delta \text{ shellable} \Rightarrow \Delta \text{ sequentially Cohen-Macaulay.}$$

Example 4.13 The simplicial complex Δ in Fig. 4.1 is a nonpure simplicial complex of dimension 2. It has 3 facets; the facet $\{a, b, c\}$ is of dimension 2, the facet $\{c, d\}$ is of dimension 1, and the facet $\{e\}$ is of dimension 0. The complex Δ is vertex decomposable with $\{e, d\}$ as a shedding order.

4.2.2 Hypergraphs

Hypergraphs are a generalization of graphs that where introduced in Chap. 2. We now introduce this combinatorial object; note that some of graph theoretic terms introduced in Chap. 2 have a hypergraph analog.

A hypergraph $H = (X, \mathscr{E})$ over the vertex set $X = \{x_1, \ldots, x_n\}$ consists of X and a collection $\mathscr{E}$ of nonempty subsets of X; these subsets are called the *edges* of H. A hypergraph H is *simple* if there is no nontrivial containment between any pair of its edges. Simple hypergraphs are also referred to as *clutters* or *Sperner systems*. All hypergraphs we consider will be simple.

When working with a hypergraph H, we shall use $X(H)$ and $\mathscr{E}(H)$ to denote its vertex and edge sets, respectively. We shall assume that hypergraphs under consideration have no *isolated vertices*, those are vertices that do not belong to any edge. An edge $\{v\}$ consisting of a single vertex is often referred to as an *isolated loop* (this is not to be confused with an isolated vertex).

Let $Y \subseteq X$ be a subset of the vertices in H. The *induced subhypergraph* of H on Y, denoted by $H[Y]$, is the hypergraph with vertex set Y and edge set $\{E \in \mathscr{E} \mid E \subseteq Y\}$. In Definition 2.23 we introduced a matching in a graph; we now extend this definition to the hypergraph context.

Definition 4.14 Let H be a simple hypergraph.

1. A collection C of edges in H is called a *matching* if the edges in C are pairwise disjoint. The maximum size of a matching in H is called its *matching number*.
2. A collection C of edges in H is called an *induced matching* if C is a matching, and C consists of all edges of the induced subhypergraph $H[\cup_{E \in C} E]$ of H. The maximum size of an induced matching in H is called its *induced matching number*.

Example 4.15 Figure 4.1 can be viewed as a hypergraph over the vertex set $V = \{a, b, c, d, e\}$ with edges $\{a, b, c\}, \{c, d\}$ and $\{e\}$. The collection $\{\{a, b, c\}, \{e\}\}$ forms an induced matching in this hypergraph.

Note that a *graph*, as introduced in Chap. 2, is a hypergraph in which all edges are of cardinality 2. We shall also need the following special family of graphs.

Definition 4.16 Let G be a simple graph on n vertices.

1. G is called *chordal* if it has no induced cycles of length ≥ 4.
2. G is called *very well-covered* if it has no isolated vertices and its minimal vertex covers all have cardinality $\dfrac{n}{2}$.

A hypergraph H is *d-uniform* if all its edges have cardinality d. For an edge E in H, let

$$N(E) = \{x \in X \mid \text{there exists } F \subseteq E \text{ such that } F \cup \{x\} \in \mathscr{E}\}$$

be the set of *neighbors* of E, and let $N[E] = N(E) \cup E$.

Definition 4.17 Let $H = (X, \mathscr{E})$ be a simple hypergraph and let E be an edge in H.

1. Define $H \setminus E$ to be the hypergraph obtained by deleting E from the edge set of H. This is often referred to as the *deletion* of E from H.
2. Define H_E to be the *contraction* of H to the vertex set $X \setminus N[E]$ (i.e., edges of H_E are minimal nonempty sets of the form $F \cap (X \setminus N[E])$, where $F \in \mathscr{E}$).

Definition 4.18 Let $H = (X, \mathscr{E})$ be a simple hypergraph.

1. A collection of vertices V in H is called an *independent set* if there is no edge $E \in \mathscr{E}$ such that $E \subseteq V$.
2. The *independence complex* of H, denoted by $\Delta(H)$, is the simplicial complex whose faces are independent sets in H.

Fig. 4.2 A simple graph
whose independence complex
is in Fig. 4.1

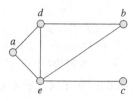

Example 4.19 The simplicial complex Δ in Fig. 4.1 is the independence complex of the graph in Fig. 4.2.

Remark 4.20 We call a hypergraph H vertex decomposable (shellable, sequentially Cohen-Macaulay) if its independence complex $\Delta(H)$ is vertex decomposable (shellable, sequentially Cohen-Macaulay).

4.2.3   Stanley-Reisner Ideals and Edge Ideals

The Stanley-Reisner ideal and edge ideal constructions are well-studied correspondences between commutative algebra and combinatorics. Those constructions arise by identifying minimal generators of a squarefree monomial ideal with the minimal nonfaces of a simplicial complex or the edges of a simple hypergraph.

Stanley-Reisner ideals were developed in the 1970s and the early 1980s (cf. [155]) and have led to many important homological results (cf. books of Bruns and Herzog [25] and Peeva [146]).

Definition 4.21 Let Δ be a simplicial complex on X. The *Stanley-Reisner ideal* of Δ is defined to be

$$I_\Delta = \langle x^F \mid F \subseteq X \text{ is not a face of } \Delta \rangle.$$

Example 4.22 Let Δ be the simplicial complex in Fig. 4.1, and we set $R = \mathbb{K}[a, b, c, d, e]$. Then the minimal nonfaces of Δ are $\{a, d\}, \{a, e\}, \{b, d\}, \{b, e\}, \{c, e\}$ and $\{d, e\}$. Thus,

$$I_\Delta = \langle ad, ae, bd, be, ce, de \rangle.$$

Example 4.23 The simplicial complex Δ in Fig. 4.3 represents a minimal triangulation of the real projective plane. Its Stanley-Reisner ideal is

$$I_\Delta = \langle abc, abe, acf, ade, adf, bcd, bdf, bef, cde, cef \rangle.$$

The edge ideal construction for hypergraphs (first studied by Hà and Van Tuyl [94]) generalizes that of graphs (already presented in Definition 2.10). This construction is similar to that of facet ideals of Faridi [71].

Fig. 4.3 A minimal triangulation of the real projective plane

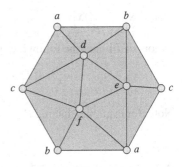

Definition 4.24 Let H be a simple hypergraph on X. The *edge ideal* of H is defined to be

$$I(H) = \langle x^E \mid E \subseteq X \text{ is an edge in } H \rangle.$$

The notions of a Stanley-Reisner ideal and an edge ideal give the following one-to-one correspondences that allow us to pass back and forth from squarefree monomial ideals to simplicial complexes and simple hypergraphs.

$$\left\{ \begin{array}{c} \text{simplicial complexes} \\ \text{over } X \end{array} \right\} \longleftrightarrow \left\{ \begin{array}{c} \text{squarefree monomial} \\ \text{ideals in } R \end{array} \right\} \longleftrightarrow \left\{ \begin{array}{c} \text{simple hypergraphs} \\ \text{over } X \end{array} \right\}.$$

In fact, every edge ideal is a Stanley-Reisner ideal and vice-versa via the notion of the independence complex. The following lemma follows directly from the definition of independence complexes and the construction of Stanley-Reisner and edge ideals.

Lemma 4.25 *Let H be a simple hypergraph and let $\Delta = \Delta(H)$ be its independence complex. Then*

$$I_\Delta = I(H).$$

Example 4.26 The edge ideal of the graph G in Fig. 4.2 is the same as the Stanley-Reisner ideal of its independence complex, the simplicial complex in Fig. 4.1.

Remark 4.27 For simplicity, if $I = I_\Delta$, then we sometimes write reg Δ for reg I, and if $I = I(H)$, then we write reg H for reg I.

For a monomial ideal in general one can pass to a squarefree monomial ideal via the polarization and still keep many important properties and invariants. We shall briefly recall the notion of polarization; see Herzog and Hibi [106] for more details.

Definition 4.28 Let $I \subseteq R = \mathbb{K}[x_1, \ldots, x_n]$ be a monomial ideal. For each $i = 1, \ldots, n$ let a_i be the maximum power of x_i appearing in the monomial generators of I. The *polarization* of I, denoted by I^{pol}, is constructed as follows.

- Let $R^{\mathrm{pol}} = \mathbb{K}[x_{11}, \ldots, x_{1a_1}, \ldots, x_{n1}, \ldots, x_{na_n}]$.
- The ideal I^{pol} is generated by monomials in R^{pol} that are obtained from generators of I under the following substitution, for each $(\gamma_1, \ldots, \gamma_n) \le (a_1, \ldots, a_n)$,

$$x_1^{\gamma_1} \ldots x_n^{\gamma_n} \longrightarrow x_{11} \ldots x_{1\gamma_1} \ldots x_{n1} \ldots x_{n\gamma_n}.$$

Note, for example, that $\operatorname{reg} R/I = \operatorname{reg} R^{\mathrm{pol}}/I^{\mathrm{pol}}$.

4.3   Hochster's and Takayama's Formulas

Hochster's and Takayama's formulas allow us to relate (multi)graded Betti numbers of a monomial ideal to the dimension of reduced homology groups of simplicial complexes. Hochster's formula deals specifically with squarefree monomial ideals, which are reflected in the next two lemmas, while Takayama's formula works for an arbitrary monomial ideal and is given later on.

The polynomial ring $R = \mathbb{K}[x_1, \ldots, x_n]$ has a natural $\mathbb{N}^n$-graded structure, and for any *monomial* ideal $I \subset R$, the quotient ring R/I inherits this $\mathbb{N}^n$-graded structure from that of R. Therefore, the torsion $\operatorname{Tor}_i^R(I, \mathbb{K})$ and the local cohomology module $H_{\mathrm{m}}^i(R/I)$ has a $\mathbb{Z}^n$-graded structure. Let $[1, n]$ denote the set $\{1, \ldots, n\}$. For $\mathbf{a} = (a_1, \ldots, a_n) \in \mathbb{Z}^n$, set $x^{\mathbf{a}} = x_1^{a_1} \cdots x_n^{a_n}$. For a monomial $\mathbf{m}$ in R, by abusing notation, we view degree $\mathbf{m}$ component of a $\mathbb{Z}^n$-graded R-module as its degree $\operatorname{supp} \mathbf{m}$ component. We shall introduce Hochster's formula following, for example, [106, Theorem 8.1.1].

Lemma 4.29 (Hochster's Formula)   *Let Δ be a simplicial complex on the vertex set $X = \{x_1, \ldots, x_n\}$ and let $\mathbf{m}$ be a monomial of R. Then,*

$$\dim_{\mathbb{K}} \operatorname{Tor}_i^R(I_\Delta, \mathbb{K})_{\mathbf{m}} = \begin{cases} \dim_{\mathbb{K}} \widetilde{H}^{\deg(\mathbf{m})-i-2}(\Delta[\operatorname{supp} \mathbf{m}]; \mathbb{K}) & \text{if } \mathbf{m} \text{ is squarefree} \\ 0 & \text{otherwise.} \end{cases}$$

In particular,

$$\beta_{ij}(I_\Delta) = \sum_{\substack{\deg(\mathbf{m}) = j, \\ \mathbf{m} \text{ is squarefree}}} \dim_{\mathbb{K}} \widetilde{H}^{j-i-2}(\Delta[\operatorname{supp} \mathbf{m}]; \mathbb{K}) \text{ for all } i, j \ge 0.$$

Here, $\Delta[\operatorname{supp} \mathbf{m}]$ is the induced subcomplex of Δ on the support of $\mathbf{m}$.

Proof We shall outline the proof of Hochster's formula following that given by Herzog and Hibi [106, Theorem 8.1.1].

1. Let $\mathscr{K}$ be the Koszul complex of I_Δ with respect to the variables $x = \{x_1, \ldots, x_n\}$, let K_i be the i-th module, and let $H_i(\mathscr{K})$ be the i-th homology

group of $\mathcal{K}$. Since $\mathcal{K}$ is a complex of $\mathbb{Z}^n$-graded modules, $H_i(\mathcal{K})$ is also a $\mathbb{Z}^n$-graded $\mathbb{K}$-vector space. Thus, for a monomial $\mathbf{m}$ in R, we have

$$\mathrm{Tor}_i^R(I_\Delta, \mathbb{K})_{\mathbf{m}} = H_i(\mathcal{K})_{\mathbf{m}}.$$

2. For $F = \{j_0 < \cdots < j_i\} \subseteq [1, n]$, set $\mathbf{e}_F = e_{j_0} \wedge \cdots \wedge e_{j_i}$. The elements $\mathbf{e}_F$'s with $|F| = i$ form a basis for the i-th free module in the Koszul complex of R with respect to x. The $\mathbb{Z}^n$-degree of $\mathbf{e}_F$ is $\epsilon(F) \in \mathbb{Z}^n$, where $\epsilon(F)$ is the $(0,1)$-vector with support F.

3. A $\mathbb{K}$-basis for $(K_i)_{\mathbf{m}}$ is given by

$$x^{\mathbf{b}} \mathbf{e}_F, \quad \text{where } \mathbf{b} + \epsilon(F) = \mathbf{m} \text{ and } \mathrm{supp}\,\mathbf{b} \notin \Delta.$$

4. Define the simplicial complex

$$\Delta^{\mathbf{m}} = \left\{ F \subseteq [1, n] \mid F \subseteq \mathrm{supp}\,\mathbf{m}, \, \mathrm{supp}\,\frac{\mathbf{m}}{x^{\epsilon(F)}} \notin \Delta \right\}.$$

Let $\widetilde{\mathcal{C}}(\Delta^{\mathbf{m}})[-1]$ be the oriented augmented chain complex of $\Delta^{\mathbf{m}}$ shifted by -1 in homological degree. Then, we have an isomorphism of complexes

$$\widetilde{\mathcal{C}}(\Delta^{\mathbf{m}})[-1] \longrightarrow \mathcal{K}_{\mathbf{m}}$$

obtained by $F = [j_0, \ldots, j_{i-2}] \mapsto \dfrac{\mathbf{m}}{x^{\epsilon(F)}} \mathbf{e}_F$. This, in turn, gives

$$H_i(\mathcal{K})_{\mathbf{m}} \simeq H_i(\widetilde{\mathcal{C}}(\Delta^{\mathbf{m}})[-1]).$$

5. If $\mathbf{m}$ is not squarefree, then there exists j such that x_j appears with power greater than 1 in $\mathbf{m}$. Define $\mathbf{m}(r) = \mathbf{m}x_j^r$ for $r \in \mathbb{N}$. It is easy to see that $\Delta^{\mathbf{m}} = \Delta^{\mathbf{m}(r)}$ for all $r \in \mathbb{N}$. Moreover, for $r \gg 0$, $H_i(\mathcal{K})_{\mathbf{m}(r)} = 0$. Thus,

$$H_i(\mathcal{K})_{\mathbf{m}} \simeq H_i(\widetilde{\mathcal{C}}(\Delta^{\mathbf{m}})[-1]) = H_i(\widetilde{\mathcal{C}}(\Delta^{\mathbf{m}(r)})[-1]) = H_i(\mathcal{K})_{\mathbf{m}(r)} = 0.$$

6. Suppose that $\mathbf{m}$ is squarefree. It can be seen that $F \subseteq \Delta^{\mathbf{m}}$ if and only if $F \subseteq \mathrm{supp}\,\mathbf{m}$ and $\mathrm{supp}\,\mathbf{m} \setminus F \notin \Delta[\mathrm{supp}\,\mathbf{m}]$. That is, $\Delta^{\mathbf{m}} = \Delta[\mathrm{supp}\,\mathbf{m}]^{\vee}$ where $(-)^{\vee}$ denotes the Alexander dual of a simplicial complex. Hence, we have

$$H_i(\widetilde{\mathcal{C}}(\Delta^{\mathbf{m}})[-1]) \simeq \widetilde{H}_{i-1}(\Delta[\mathrm{supp}\,\mathbf{m}]^{\vee}; \mathbb{K}) \simeq \widetilde{H}^{\deg\mathbf{m}-i-2}(\Delta[\mathrm{supp}\,\mathbf{m}]; \mathbb{K})$$

where the second isomorphism is a standard fact about Alexander duality.

Lemma 4.30 *For a simplicial complex Δ, the following are equivalent:*

1. $\operatorname{reg} R/I_\Delta \geq d$.
2. $\tilde{H}_{d-1}(\Delta[S], \mathbb{K}) \neq 0$, *where $\Delta[S]$ denotes the induced subcomplex on some subset S of vertices.*
3. $\tilde{H}_{d-1}(\operatorname{link}_\Delta \sigma, \mathbb{K}) \neq 0$ *for some face σ of Δ.*

Proof The equivalence of (1) and (2) follows directly from Definition 4.1, together with Hochster's formula in Lemma 4.29. The equivalence of (1) and (3) follows directly from the local cohomology characterization of regularity, together with the fact that $H_{\mathfrak{m}}^i(R/I_\Delta, \mathbb{K})_{-\sigma} \simeq \tilde{H}^{i-|\sigma|-1}(\operatorname{link}_\Delta \sigma, \mathbb{K})$ (see Miller and Sturmfels book [137, Chapter 13.2]).

We will also make use of a variation of Hochster's formula following [137, Theorem 1.34]. This variation of Hochster's formula is given via upper-Koszul simplicial complexes associated to monomial ideals.

Definition 4.31 Let $I \subseteq R$ be a monomial ideal and let $\boldsymbol{\alpha} = (\alpha_1, \ldots, \alpha_n) \in \mathbb{N}^n$ be a $\mathbb{N}^n$-graded degree. The *upper-Koszul simplicial complex* associated to I at degree $\boldsymbol{\alpha}$, denoted by $K^{\boldsymbol{\alpha}}(I)$, is the simplicial complex over $X = \{x_1, \ldots, x_n\}$ whose faces are:

$$\left\{ W \subseteq X \ \middle| \ \frac{x^{\boldsymbol{\alpha}}}{\prod\limits_{u \in W} u} \in I \right\}.$$

Theorem 4.32 ([137, Theorem 1.34]) *Let $I \subseteq R$ be a monomial ideal. Then its $\mathbb{N}^n$-graded Betti numbers are given as follows:*

$$\beta_{i,\boldsymbol{\alpha}}(I) = \dim_{\mathbb{K}} \tilde{H}_{i-1}(K^{\boldsymbol{\alpha}}(I); \mathbb{K}) \text{ for } i \geq 0 \text{ and } \boldsymbol{\alpha} \in \mathbb{N}^n. \tag{4.1}$$

Takayama's formula [160, Theorem 1] describes the the dimension of the $\mathbb{Z}^n$-graded component $H_{\mathfrak{m}}^i(R/I)_{\mathbf{a}}$, for $\mathbf{a} \in \mathbb{Z}^n$, in terms of a simplicial complex $\Delta_{\mathbf{a}}(I)$. We shall recall the construction of $\Delta_{\mathbf{a}}(I)$, as given by Minh and Trung [138], which is simpler than the original construction of [160].

For $\mathbf{a} = (a_1, \ldots, a_n) \in \mathbb{Z}^n$, set $G_{\mathbf{a}} := \{j \in [1, n] \mid a_j < 0\}$. For every subset $F \subseteq [1, n]$, let $R_F = R[x_j^{-1} \mid j \in F]$. Define

$$\Delta_{\mathbf{a}}(I) = \{F \setminus G_{\mathbf{a}} \mid G_{\mathbf{a}} \subseteq F, \ x^{\mathbf{a}} \notin I R_F\}.$$

We call $\Delta_{\mathbf{a}}(I)$ a *degree complex* of I.

Lemma 4.33 (Takayama's Formula) *For any $\mathbf{a} \in \mathbb{Z}^n$, we have*

$$\dim_{\mathbb{K}} H_{\mathfrak{m}}^i(R/I)_{\mathbf{a}} = \dim_{\mathbb{K}} \tilde{H}_{i-|G_{\mathbf{a}}|-1}(\Delta_{\mathbf{a}}(I), \mathbb{K}).$$

The original formula in [160, Theorem 1] is slightly different. It contains additional conditions on $\mathbf{a}$ for $H_{\mathfrak{m}}^i(R/I)_{\mathbf{a}} = 0$. However, the proof in [160] shows that we may drop these conditions, which is more convenient for our investigation.

From Takayama's formula we immediately obtain the following characterizations of depth and regularity of monomial ideals in terms of the degree complexes.

Lemma 4.34 *Let $I \subseteq R$ be a monomial ideal. Then*

$$\operatorname{reg} R/I = \max\{|\mathbf{a}| + |G_{\mathbf{a}}| + i \mid \mathbf{a} \in \mathbb{Z}^n, i \geq 0, \widetilde{H}_{i-1}(\Delta_{\mathbf{a}}(I), \mathbb{K}) \neq 0\}.$$

Chapter 5
Problems, Questions, and Inductive Techniques

In this chapter, we present a number of open problems and questions for edge ideals of graphs. These problems and questions fall under the umbrella of Problem 4.8. We shall also discuss inductive techniques that have been applied in the literature.

5.1 Regularity of Powers of Edge Ideals

When restricted to the case that $I = I(G)$ is the edge ideal of a simple graph G, it is known by Theorem 4.7 that for $q \gg 0$,

$$\operatorname{reg} I^q = 2q + b$$

and Problem 4.8 is to determine the constants b and $q_0 = \min\{t \mid \operatorname{reg} I^q = 2q + b$ for all $q \geq t\}$. This has been accomplished for a number of special classes of graphs, namely, for forests, cycles, unicyclic graphs, and very well-covered graphs, and is due to the work of Beyarslan, Hà, and Trung [15], Alilooee, Beyarslan, and Selvaraja [3], Moghimian, Seyed Fakhari, and Yassemi [139], and Norouzi, Seyed Fakhari, and Yassemi [144].

Recall from Definition 4.14 that an induced matching C of a graph G is a matching C such that induced subgraph of G on the vertices of C is the matching C. The induced matching number, denoted $v(G)$, is the size of the maximum induced matching.

Theorem 5.1 ([15, Theorem 4.7]) *Let G be a forest, and let $I = I(G)$ be its edge ideal. Then for all $q \geq 1$, we have*

$$\operatorname{reg} I^q = 2q + v(G) - 1.$$

© The Editor(s) (if applicable) and The Author(s), under exclusive
licence to Springer Nature Switzerland AG 2020
E. Carlini et al., *Ideals of Powers and Powers of Ideals*, Lecture Notes of the Unione
Matematica Italiana 27, https://doi.org/10.1007/978-3-030-45247-6_5

Theorem 5.2 ([15, Theorem 5.2])   *Let C_n denote the n-cycle, and let $I = I(C_n)$ be its edge ideal. Let $v = \lfloor \frac{n}{3} \rfloor$ be the induced matching number of C_n. Then*

$$\operatorname{reg} I = \begin{cases} v + 1 \text{ if } n \equiv 0, 1 \ (mod\ 3) \\ v + 2 \text{ if } n \equiv 2 \quad (mod\ 3) \end{cases}$$

and for any $q \geq 2$, we have

$$\operatorname{reg} I^q = 2q + v - 1.$$

Theorem 5.3 ([3, Theorem 1.2], [139, Proposition 1.1])   *Let G be a unicyclic graph (i.e., a graph having exactly one cycle) that is not a cycle, and let $I = I(G)$ be its edge ideal. Then for all $q \geq 1$, we have*

$$\operatorname{reg} I^q = 2q + \operatorname{reg} I - 2.$$

Theorem 5.4 ([120, Theorem 5.3], [144, Theorem 3.6])   *Let G be a very well-covered graph, and let $I = I(G)$ be its edge ideal. Then for all $q \geq 1$, we have*

$$\operatorname{reg} I^q = 2q + v(G) - 1.$$

These theorems give rise to the following problem.

Problem 5.5   Characterize graphs G for which the edge ideals $I = I(G)$ satisfy

1. $\operatorname{reg} I^q = 2q + v(G) - 1$ for all $q \gg 0$.
2. $\operatorname{reg} I^q = 2q + \operatorname{reg} I - 2$ for all $q \gg 0$.

The simplest situation for an edge ideal is when its powers have linear resolutions. It is a nice result (see Fröberg [78] and Wegner [169]) that the edge ideal of a graph G has a linear resolution if and only if G^c is chordal. It also follows from work of Herzog, Hibi, and Zheng [110] that if $I(G)$ has a linear resolution, then so does $I(G)^q$ for all $q \geq 1$. It is, thus, of interest to characterize graphs whose (sufficiently large) powers have linear resolutions. It is known (see Nevo and Peeva [143]) that if a power of $I(G)$ has a linear resolution, then G^c has no induced 4-cycles.

Problem 5.6 (Francisco-Hà-Van Tuyl and Nevo-Peeva)   Suppose that $v(G) = 1$, i.e., G^c has no induced 4-cycle and let $I = I(G)$.

1. Prove (or disprove) that $\operatorname{reg} I^q = 2q$ for all $q \gg 0$.
2. Prove (or disprove) that $\operatorname{reg} I^{q+1} = \operatorname{reg} I^q + 2$ for all $q \geq \operatorname{reg} I - 1$.

A *cricket* graph is obtained by attaching exactly two leaves at the same vertex of a triangle (a C_3). A *diamond* graph is obtained by connecting exactly one pair of opposite vertices in a C_4. A graph G is called *cricket-free* or *diamond-free* if G has no induced subgraph that is a cricket or a diamond, respectively. Problem 5.6 has an affirmative answer (for all $q \geq 2$) under an additional condition that G is also cricket-free or diamond-free Banerjee [7] and Erey [68].

Computational experiments also suggest that the constant q_0, i.e., the first place where reg I^q becomes a linear function, cannot be too big.

Question 5.7 Is it true that $q_0 \leq$ reg $I(G)$?

It is often difficult to get the exact value for the asymptotic linear function reg I^q. Linear bounds are also of interest. The following general lower bound of Beyarslan, Hà, and Trung [15] was inspired by a result of Katzman [125], who proved the bound when $q = 1$ (i.e., for the edge ideal itself).

Theorem 5.8 ([15, Theorem 4.5]) *Let G be any graph, and let $I = I(G)$ be its edge ideal. Then for all $q \geq 1$, we have*

$$\text{reg } I^q \geq 2q + \nu(G) - 1.$$

A lower bound would be especially interesting when coupling with an upper bound. Unfortunately, there has not been any satisfactory general upper bound for reg I^q. The following conjecture seems plausible.

Conjecture 5.9 (Banerjee, Beyarslan and Hà) Let G be any graph, and let $I = I(G)$ be its edge ideal. Then for all $q \geq 1$, we have

$$\text{reg } I^q \leq 2q + \text{reg } I - 2.$$

For bipartite graphs, a slightly weaker upper bound was obtained by Jayanthan, Narayanan and Selvaraja [122]. Recall that co-chord(G), the *co-chordal number* of G, denotes the least number of co-chordal subgraphs (graphs whose complements are chordal) of G whose union is G. It was proved by Woodroofe [171] that

$$\text{reg } I(G) \leq \text{co-chord}(G) + 1.$$

An upper bound for reg I^q for *bipartite* graphs is stated as follows.

Theorem 5.10 ([122, Theorem 1.1]) *Let G be a bipartite graph, and let $I = I(G)$ be its edge ideal. Then for all $q \geq 1$, we have*

$$\text{reg } I^q \leq 2q + \text{co-chord}(G) - 1.$$

Inspired by Theorem 5.10, one could might want to either attack Conjecture 5.9 for bipartite graphs or get a similar bound as that of Theorem 5.10 for all graphs.

5.2   Regularity of Symbolic Powers of Edge Ideals

Symbolic powers of squarefree monomial ideals (particularly, edge ideals of graphs) are quite easy to describe. Let G be a simple graph, and let $I = I(G)$ be its edge ideal. Then by Lemma 2.13 the ideal I has the following primary decomposition:

$$I = \bigcap_{W \text{ is a minimal vertex cover}} \langle x \mid x \in W \rangle.$$

In this case, the *symbolic powers of I* are given in the following way: for all $q \geq 1$,

$$I^{(q)} = \bigcap_{W \text{ is a minimal vertex cover}} \langle x \mid x \in W \rangle^q.$$

The study of the regularity of symbolic powers is much more subtle than that of ordinary powers. It is known by Herzog, Hoa, and Trung [109] that the regularity of symbolic powers of an edge ideal (or a monomial ideal in general) is bounded above by a linear function. However, this linear function is often too big to give exact values. The following natural question remains open.

Question 5.11 Let G be a simple graph, and let $I = I(G)$ be its edge ideal. Is reg $I^{(q)}$ asymptotically a linear function, i.e., are there constants a and b such that

$$\text{reg } I^{(q)} = aq + b \text{ for all } q \gg 0?$$

A bolder statement than Question 5.11 to investigate is the following question.

Question 5.12 (N.C Minh)   Let G be a graph, and let $I = I(G)$ be its edge ideal. Is reg $I^{(q)} = \text{reg } I^q$ for all $q \gg 0$?

This question is certainly not true for squarefree monomial ideals in general. A good place to start investigating Question 5.12 is the following problem.

Question 5.13 Let G be a co-chordal graph (i.e., G^c is chordal). Prove (or disprove) that for all $q \gg 0$, we have

$$\text{reg } I^{(q)} = 2q.$$

As in the case for ordinary powers, general lower and upper bounds for reg $I^{(q)}$ are also of interest. To this end, we ask if a similar bound to that of Theorem 5.8 holds also for symbolic powers.

Question 5.14 Let G be a graph, and let $I = I(G)$ be its edge ideal. Is it true that, for all $q \geq 1$,

$$\text{reg } I^{(q)} \geq 2q + \nu(G) - 1?$$

This question also leads to the following problem.

Problem 5.15

1. Find classes of graphs for which reg $I^{(q)} = 2q + \nu(G) - 1$ for all $q \geq 1$.
2. Find a *good* linear upper bound for reg $I^{(q)}$ for $q \gg 0$.

The first difficulty when working with Questions 5.11–5.15 is how to understand symbolic powers of edge ideals. One might try Questions 5.11–5.15 for special classes of graphs for which the symbolic powers of their edge ideals are well described. The class of *perfect* graphs is a good place to start due to the following result of Sullivant [157]. For a graph G, let $C_2(G)$ denote the set of cliques of size at least 2 in G.

Theorem 5.16 ([157, Theorem 3.10]) *A graph G is perfect if and only if for all $q \geq 1$, we have*

$$I(G)^{(q)} = \left\langle \prod_{i=1}^{l} x^{V_i} \,\middle|\, G[V_i] \in C_2(G) \text{ with } \sum_{i=1}^{l}(|V_i| - 1) = q \right\rangle.$$

One can also consider small symbolic powers of edge ideals. For instance, the second symbolic power of the edge ideal of a graph is quite well understood.

Theorem 5.17 ([157, Corollary 3.12]) *Let G be any graph, and let $I = I(G)$ be its edge ideal. Then $I^{(2)}$ is generated by cubics of the form $x_i x_j x_k$, where $\{x_i, x_j, x_k\}$ is a triangle in G, and quartics of the form $x_i x_j x_k x_l$, where $\{x_i, x_j\}$ and $\{x_k, x_l\}$ are edges in G.*

5.3 Inductive Techniques

The backbone of most of the obtained results on the regularity of powers of edge ideals is mathematical induction. The underlying idea is to relate the regularity of powers of a squarefree monomial ideal, corresponding to a simplicial complex and/or hypergraph, to that of smaller ideals, corresponding to subcomplexes and/or subhypergraphs. In this section we recall a number of inductive results that played the key rôle in most of these studies.

We start with a few crude bounds for the regularity of hypergraphs and simplicial complexes. These bounds follow immediately from Lemma 4.30.

Lemma 5.18

1. *Let H be a simple hypergraph. Then* reg $H \geq$ reg H' *for any induced subhypergraph H' of H.*
2. *Let Δ be a simplicial complex. Then* reg $\Delta \geq$ reg $\text{link}_\Delta(\sigma)$ *for any face σ of Δ.*

Next, we recall a number of bounds for the regularity of hypergraphs and simplicial complexes that result from familiar short exact sequences. In particular, for homogeneous ideals $I, J \subseteq R$ and a homogeneous element $h \in R$, the following short exact sequences are standard in commutative algebra.

$$0 \longrightarrow \frac{R}{I:h}(-d) \xrightarrow{\times h} \frac{R}{I} \longrightarrow \frac{R}{I+h} \longrightarrow 0, \quad \text{and} \tag{5.1}$$

$$0 \longrightarrow \frac{R}{I \cap J} \longrightarrow \frac{R}{I} \oplus \frac{R}{J} \longrightarrow \frac{R}{I+J} \longrightarrow 0. \tag{5.2}$$

The long exact sequence of local cohomology modules associated to (5.1) gives us a simple bound:

$$\operatorname{reg} I \le \max\{\operatorname{reg}(I:h)+d, \operatorname{reg}(I,h)\}. \tag{5.3}$$

Remark 5.19 If I is a monomial ideal and h is an indeterminate of R appearing in the generators of I, then it was shown by Dao, Huneke, and Schweig [49, Lemma 2.10] that, in fact, $\operatorname{reg} I$ is always equal to either $\operatorname{reg}(I:h)+1$ or $\operatorname{reg}(I,h)$.

In practice, induction based on the combinatorial structures of hypergraphs and simplicial complexes is often performed by successively deleting a vertex or an edge (or a face). In these situations, h is either a variable or a product of variables on an edge (or a face).

Remark 5.20 Let $I \subseteq R$ be a squarefree monomial ideal.

1. Suppose that $I = I_\Delta$ for a simplicial complex Δ. Let $h = x^\sigma$, where $\sigma \in \Delta$ is of dimension $d-1$. Then $I : \langle h \rangle = I_{\operatorname{link}_\Delta(\sigma)}$ and $I + \langle h \rangle = I_{\operatorname{del}_\Delta(\sigma)} + \langle h \rangle$.
2. Suppose that $I = I(H)$ is the edge ideal of a simple hypergraph H. For a subset V of the vertices, $|V| = d$, let $H : V$ and $H + V$ represent the simple hypergraphs corresponding to the squarefree monomial ideals $I(H) : \langle x^V \rangle$ and $I(H) + \langle x^V \rangle$, respectively.

As a consequence of (5.3) and the above remark we have the following inductive bounds.

Theorem 5.21

1. Let Δ be a simplicial complex, and let σ be a face of dimension $d-1$ in Δ. Then

$$\operatorname{reg} \Delta \le \max\{\operatorname{reg} \operatorname{link}_\Delta(\sigma) + d, \operatorname{reg} \operatorname{del}_\Delta(\sigma)\}.$$

2. Let H be a simple hypergraph, and let V be a collection of d vertices in H. Then

$$\operatorname{reg} H \le \max\{\operatorname{reg}(H:V)+d, \operatorname{reg}(H+V)\}.$$

Now, let E be an edge of a simple hypergraph H. By taking $I = I(H \setminus E)$ and $J = \langle x^E \rangle$ in the short exact sequence (5.2), we have

$$0 \longrightarrow \frac{R}{\langle x^E \rangle \cap I(H \setminus E)} \longrightarrow \frac{R}{\langle x^E \rangle} \oplus \frac{R}{I(H \setminus E)} \longrightarrow \frac{R}{I(H)} \longrightarrow 0.$$

Taking the associated long exact sequence of cohomology modules again, we get

$$\operatorname{reg} H \leq \max\{|E|, \operatorname{reg}(H \setminus E), \operatorname{reg}\left(\langle x^E \rangle \cap I(H \setminus E)\right) - 1\}. \tag{5.4}$$

Observe that $\langle x^E \rangle \cap I(H \setminus E) = x^E \langle y \mid y \in N(E) \rangle + I(H_E)$, where H_E denotes the contraction of $H \setminus N(E)$ to the vertices $X \setminus N[E]$ (i.e., edges of H_E are minimal nonempty elements of $\{F \cap (X \setminus N[E]) \mid F \in \mathscr{E}(H)\}$). Moreover, since vertex set of H_E is disjoint from $N[E]$, by taking the tensor product of minimal free resolutions, we get

$$\operatorname{reg}\left(\langle x^E \rangle \cap I(H \setminus E)\right) = \operatorname{reg}(I(H_E)) + |E|.$$

Thus, (5.4) gives the following inductive bound.

Theorem 5.22 *Let H be a simple hypergraph, and let E be an edge of cardinality d in H. Then*

$$\operatorname{reg} H \leq \max\{d, \operatorname{reg}(H \setminus E), \operatorname{reg}(H_E) + d - 1\}.$$

Another inductive bound, which does not rely on short exact sequences (5.1) and (5.2), was established by Kalai and Meshulam [124]. This bound was also extended to arbitrary (not necessarily squarefree) monomial ideals by Herzog [105].

Theorem 5.23 *Let $I_1, \ldots, I_s$ be squarefree monomial ideals in R. Then*

$$\operatorname{reg}\left(R \Big/ \sum_{i=1}^{s} I_i\right) \leq \sum_{i=1}^{s} \operatorname{reg} R/I_i.$$

If we restrict to edge ideals of hypergraphs, Theorem 5.23 immediately gives us the following corollary.

Corollary 5.24 *Let H and $H_1, \ldots, H_s$ be simple hypergraphs over the same vertex set X such that $\mathscr{E}(H) = \bigcup_{i=1}^{s} \mathscr{E}(H_i)$. Then*

$$\operatorname{reg} R/I(H) \leq \sum_{i=1}^{s} \operatorname{reg} R/I(H_i).$$

When working with powers of edge ideals, induction not only goes from a given hypergraph to subhypergraphs, but also from a bigger power to smaller ones. Banerjee's recent work [7] facilitates this inductive process. Banerjee's technique has proved to be quite powerful in getting the exact linear form for the regularity of powers of edge ideals of special classes of graphs. We recall the following definition and theorems from [7].

Fig. 5.1 A graph and
even-connected vertices

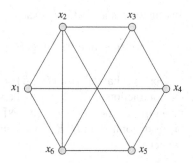

Theorem 5.25 (see [7, Theorem 5.2]) *Let G be a graph, and let s be a positive integer. Denote the set of minimal monomial generators of $I(G)^s$ by $\{m_1, \ldots, m_k\}$. Then*

$$\operatorname{reg} I(G)^{s+1} \leq \max\{\operatorname{reg} I(G)^s, \operatorname{reg}\left(I(G)^{s+1} : \langle m_l \rangle\right) + 2s, 1 \leq l \leq k\}.$$

Definition 5.26 Let $G = (V, E)$ be a graph. Two vertices u and v in G are said to be *even-connected* with respect to an s-fold product $M = x^{e_1} \cdots x^{e_s}$, where $e_1, \ldots, e_s$ are edges in G, if there is a path $p_0, \ldots, p_{2l+1}$, for some $l \geq 1$, in G such that the following conditions hold:

1. $p_0 \equiv u$ and $p_{2l+1} \equiv v$;
2. for all $0 \leq j \leq l - 1$, $\{p_{2j+1}, p_{2j+2}\} = e_i$ for some i; and
3. for all i, $\left|\{j \mid \{p_{2j+1}, p_{2j+2}\} = e_i\}\right| \leq \left|\{t \mid e_t = e_i\}\right|$.

Theorem 5.27 ([7, Theorems 6.1 and 6.7]) *Let $G = (V, E)$ be a graph with edge ideal $I = I(G)$, and let $s \geq 1$ be an integer. Let $M = x^{e_1} \cdots x^{e_s}$ be a minimal generator of I^s. Then $I^{s+1} : \langle M \rangle$ is minimally generated by monomials of degree 2, and uv (u and v may be the same) is a minimal generator of $I^{s+1} : \langle M \rangle$ if and only if either $\{u, v\} \in E$ or u and v are even-connected with respect to M.*

Example 5.28 Let G be the graph in Fig. 5.1, and let $I = I(G) \subseteq k[x_1, \ldots, x_6]$. Let $e = \{x_2, x_6\}$. Then x_3 and x_5 is even-connected with respect to $M = x^e = x_2 x_6$. The path $p_0, \ldots, p_3$ as in Definition 5.26 can be chosen to be x_3, x_2, x_6, x_5. In particular, $x_3 x_5 \in I^2 : \langle M \rangle$ by Theorem 5.27.

Chapter 6
Examples of the Inductive Techniques

In this chapter, we present detailed proofs of a few stated results to illustrate how the inductive techniques introduced in the last chapter can be applied to the study of the regularity of powers of edge ideals.

In the first section, we examine the proof of Theorem 5.8. This theorem gives a general lower bound for $\operatorname{reg} I(G)^q$, $q \geq 1$, where G is an arbitrary graph. Lemma 5.18 shows $\operatorname{reg} I(G) \geq \operatorname{reg} I(H)$, for any induced subgraph H of G. The main idea behind the proof of Theorem 5.8 is to show that a similar statement holds when we consider the powers $I(G)^q$ and $I(H)^q$, and to find an induced subgraph H which attains the lower bound.

In the second section, we exhibit the proof of Theorem 5.1, which gives the exact formula for $\operatorname{reg} I(G)^q$, $q \geq 1$, when G is a forest. By means of the general lower bound, it remains to prove the upper bound. We apply an interesting non-standard induction to find this upper bound.

The last section is devoted to the proof of Theorem 5.2, which gives the exact formula for $\operatorname{reg} I(G)^q$, $q \geq 1$, when G is a cycle. This proof is an example of how Banerjee's induction method works.

6.1 Proof of Theorem 5.8

We start by generalizing the crude bound given in Lemma 5.18, making use of Theorem 4.32, to get a bound for the graded Betti numbers for any powers of an edge ideal.

Lemma 6.1 *Let G be a graph, and let H be an induced subgraph of G. Then for any $q \geq 1$ and any $i, j \geq 0$, we have*

$$\beta_{i,j}(I(H)^q) \leq \beta_{i,j}(I(G)^q).$$

E. Carlini et al., *Ideals of Powers and Powers of Ideals*, Lecture Notes of the Unione Matematica Italiana 27, https://doi.org/10.1007/978-3-030-45247-6_6

Proof For an $\mathbb{N}^n$-graded degree $\alpha = (\alpha_1, \ldots, \alpha_n)$, let $\text{supp}(\alpha) = \{x_i \mid \alpha_i \neq 0\}$ be the support of α. Observe that since H is an induced subgraph of G, if $\text{supp}(\alpha) \subseteq V_H$ then $K^\alpha(I(H)^q) = K^\alpha(I(G)^q)$. Thus, it follows from (4.1) that

$$\beta_{i,\alpha}(I(H)^q) = \dim_\mathbb{K} \widetilde{H}_{i-1}(K^\alpha(I(H)^q); \mathbb{K})$$
$$= \dim_\mathbb{K} \widetilde{H}_{i-1}(K^\alpha(I(G)^q); \mathbb{K}) = \beta_{i,\alpha}(I(G)^q).$$

Hence,

$$\beta_{i,j}(I(H)^q) = \sum_{\alpha \in \mathbb{N}^n, \ \text{supp}(\alpha) \subseteq V_H, |\alpha|=j} \beta_{i,\alpha}(I(H)^q) = \sum_{\alpha \in \mathbb{N}^n, \ \text{supp}(\alpha) \subseteq V_H, |\alpha|=j} \beta_{i,\alpha}(I(G)^q)$$

$$\leq \sum_{\alpha \in \mathbb{N}^n, \ |\alpha|=j} \beta_{i,\alpha}(I(G)^q) = \beta_{i,j}(I(G)^q).$$

Corollary 6.2 *Let G be a graph, and let H be an induced subgraph of G. Then, for all $q \geq 1$,*

$$\text{reg } I(H)^q \leq \text{reg } I(G)^q.$$

We shall also need the following lemma.

Lemma 6.3 *Let $F_1, \ldots, F_r$ be a regular sequence of homogeneous polynomials in R with $\deg F_1 = \cdots = \deg F_r = d$. Let $I = \langle F_1, \ldots, F_r \rangle$. Then for all $q \geq 1$, we have*

$$\text{reg } I^q = dq + (d-1)(r-1).$$

Proof We use induction on r. The statement is clear for $r = 1$. Suppose that $r > 1$. We proceed by induction on q. The statement is also clear if $q = 1$ by the Koszul complex. Thus, we may assume that $q \geq 2$.

Let $J = (F_1, \ldots, F_{r-1})$. Consider the following homomorphism

$$\phi : I^{q-1}(-d) \oplus J^q \xrightarrow{(F_r, 1)} I^q.$$

Since $I^q = (J + (F_r))^q = J^q + F_r I^{q-1}$, ϕ is surjective. Moreover, since F_r is regular in R/J, the kernel of ϕ is given by $F_r J^q$. Thus, we have the following short exact sequence

$$0 \longrightarrow J^q(-d) \longrightarrow I^{q-1}(-d) \oplus J^q \xrightarrow{(F_r, 1)} I^q \longrightarrow 0. \qquad (6.1)$$

By the induction hypothesis on r, we have $\text{reg}(J^q(-d)) = \text{reg } J^q + d = dq + (d-1)(r-1) + 1$. Furthermore, by the induction hypothesis on q, we have

$$\text{reg}(I^{q-1}(-d)) = \text{reg } I^{q-1} + d = dq + (d-1)(r-1).$$

Thus,

$$\operatorname{reg}(I^{q-1}(-d) \oplus J^q) = dq + (d-1)(r-1).$$

Combining with (6.1) and Lemma 4.6, we can conclude that

$$\operatorname{reg} I^q = dq + (d-1)(r-1),$$

and the lemma is proved.

We are now ready to establish the general bound for $\operatorname{reg} I(G)^q$ stated in Theorem 5.8. For simplicity of notation, let $r = \nu(G)$. Suppose that $\{u_1 v_1, \ldots, u_r v_r\}$ is an induced matching in G. Let H be the induced subgraph of G on the vertices $\bigcup_{i=1}^r \{u_i, v_i\}$. Then $I(H) = \langle u_1 v_1, \ldots, u_r v_r \rangle$ is a complete intersection. Thus, by Lemma 6.3, we have

$$\operatorname{reg} I(H)^q = 2q + (2-1)(r-1) = 2q + \nu(G) - 1.$$

It now follows from Corollary 6.2 that

$$\operatorname{reg} I(G)^q \geq \operatorname{reg} I(H)^q \geq 2q + \nu(G) - 1.$$

6.2 Proof of Theorem 5.1

The heart of the proof of Theorem 5.1 lies in the following lemma which, when taking H to be the empty graph, gives us the necessary upper bound for the conclusion of Theorem 5.1 to hold.

Lemma 6.4 *Let K be a forest, and let $\nu(K)$ be its induced matching number. Suppose that G and H are induced subgraphs of K such that*

$$E(H) \cup E(G) = E(K) \text{ and } E(H) \cap E(G) = \emptyset.$$

Then, for all $q \geq 1$, we have

$$\operatorname{reg}(I(H) + I(G)^q) \leq 2q + \nu(K) - 1.$$

Proof We shall use induction on $m := q + |V(G)|$. If $m = 1$, then we must have $q = 1$ and $V(G) = \emptyset$. In this case, $E(H) = E(K)$, $I(H) + I(G)^q = I(K)$, and the desired inequality is

$$\operatorname{reg} I(K) \leq \nu(K) + 1. \tag{6.2}$$

This is true and (6.2) is, in fact, an equality, following Zheng [173, Theorem 2.18]. The key steps to prove [173, Theorem 2.18] are summarized as follows:

1. Let x be a leaf vertex of K. By (5.3), we have

$$\operatorname{reg} I(K) \leq \max \left\{\operatorname{reg}\left(I(K):\langle x\rangle\right)+1, \operatorname{reg}\left(I(K), x\right)\right\}.$$

2. Note that $I(K):\langle x\rangle$ corresponds to the edge ideal of $K \setminus N_K[x]$ and $(I(K), x)$ corresponds to the edge ideal of $K \setminus x$.
3. It can be shown that $\nu(K \setminus N_K[x])+1 \leq \nu(K)$ and $\nu(K \setminus x) \leq \nu(K)$.
4. Equation (6.2) then follows by induction on $|V(K)|$.

Suppose now that $m \geq 2$. If G consists of no edges, then $E(H) = E(K)$ and we have $I(H) + I(G)^q = I(K)$. The assertion again follows from [173, Theorem 2.18].

Assume that $E(G) \neq \emptyset$. Being a subgraph of K, G is a forest. In particular, G contains a leaf. Let x be a leaf in G and let y be the unique neighbor of x in G. By Morey [140, Lemma 2.10], we have $I(G)^q : \langle xy \rangle = I(G)^{q-1}$. The ideas of the proof of [140, Lemma 2.10] are as follows:

1. The containment $I(G)^{q-1} \subseteq I(G)^q : \langle xy \rangle$ is straightforward.
2. Suppose that a is a monomial in $I(G)^q : \langle xy \rangle$. That is, $axy = e_1 \ldots e_q h$, where the e_i's correspond to edges in G and h is a monomial.
3. If x does not appear in $e_1 \ldots e_q$ then $a \in I(G)^{q-1}$. Otherwise, say $x \in e_i$. Since x is a leaf vertex and y is the only neighbor of x, we must have $e_i = xy$. In this case, $a = e_1 \ldots \widehat{e_i} \ldots e_q h$ is also in $I(G)^{q-1}$.

Since all ideals being discussed are monomial ideals, this implies that

$$(I(H) + I(G)^q) : \langle xy \rangle = (I(H) : \langle xy \rangle) + (I(G)^q : \langle xy \rangle) = (I(H) : \langle xy \rangle) + I(G)^{q-1}.$$

Moreover,

$$\langle xy \rangle + I(H) + I(G)^q = \langle xy \rangle + I(H) + I(G \setminus x)^q = I(H + xy) + I(G \setminus x)^q,$$

where $H + xy$ denotes the graph H adjoined with the edge $\{x, y\}$. Therefore, we have the following short exact sequence:

$$0 \to \left(R \big/ (I(H) : \langle xy \rangle + I(G)^{q-1})\right)(-2) \to R \big/ (I(H) + I(G)^q)$$
$$\to R / I(H + xy) + I(G \setminus x)^q \to 0.$$

This yields

$$\operatorname{reg}(I(H) + I(G)^q) \leq \max\{\operatorname{reg}(I(H) : \langle xy \rangle + I(G)^{q-1})$$
$$+2, \operatorname{reg}(I(H + xy) + I(G \setminus x)^q)\}. \tag{6.3}$$

Let $\{u_1, \ldots, u_p\} = N_H(x) \cup N_H(y)$ be all the vertices of H which are adjacent to either x or y. Let $H' = H \setminus \{u_1, \ldots, u_p\}$. Then,

$$I(H) : \langle xy \rangle = I(H') + \langle u_1, \ldots, u_p \rangle.$$

Observe that since $E(G) \cap E(H) = \emptyset$, none of the vertices $\{u_1, \ldots, u_p\}$ are in G. Therefore, we have

$$\mathrm{reg}(I(H) : \langle xy \rangle + I(G)^{q-1}) = \mathrm{reg}(I(H') + \langle u_1, \ldots, u_p \rangle + I(G)^{q-1})$$
$$= \mathrm{reg}(I(H') + I(G)^{q-1}).$$

This, coupled with (6.3), implies that

$$\mathrm{reg}(I(H) + I(G)^q) \leq \max\{\mathrm{reg}(I(H') + I(G)^{q-1})$$
$$+ 2, \mathrm{reg}(I(H + xy) + I(G \setminus x)^q)\}. \qquad (6.4)$$

Let $K' := H' \cup G$. Since $E(H) \cap E(G) = \emptyset$, we have $E(H') \cap E(G) = \emptyset$. This implies that K' is an induced subgraph of K. Thus, K' is a forest, and

$$\nu(K') \leq \nu(K).$$

Now, applying the induction hypothesis to K', G and H' with power $(q - 1)$, we have

$$\mathrm{reg}(I(H') + I(G)^{q-1}) \leq 2(q - 1) + \nu(K') - 1 \leq 2(q - 1) + \nu(K) - 1. \qquad (6.5)$$

On the other hand, since x is a leaf of G, we have $E(H + xy) \cap E(G \setminus x) = \emptyset$ and $K = (H + xy) \cup (G \setminus x)$. Thus, we can apply the induction hypothesis to $K, G \setminus x$ and $H + xy$ to get

$$\mathrm{reg}(I(H + xy) + I(G \setminus x)^q) \leq 2q + \nu(K) - 1. \qquad (6.6)$$

Putting (6.4)–(6.6) together we get the desired inequality

$$\mathrm{reg}(I(H) + I(G)^q) \leq 2q + \nu(K) - 1,$$

and the lemma is proved.

We are now ready to present the proof of Theorem 5.1. Let G be a forest, and let $I = I(G)$. For any $q \geq 1$, by Theorem 5.8 we have $\mathrm{reg}\, I^q \geq 2q + \nu(G) - 1$. On the other hand, applying Lemma 6.4 by taking $K = G$ and $H = \emptyset$, we get reg $I^q \leq 2q + \nu(G) - 1$. Hence, for all $q \geq 1$,

$$\text{reg } I^q = 2q + \nu(G) - 1.$$

6.3 Proof of Theorem 5.2

The proof of Theorem 5.2 makes use of Banerjee's inductive method. To start we shall see how Theorem 5.27 applies to our situation, where the given graph is a cycle.

Lemma 6.5 *Let C_n be the n-cycle and assume that its vertices (in order) are $x_1, \ldots, x_n$. Let $I = I(C_n)$. Then*

$$x_n^2 \in (I^{q+1} : \langle M \rangle),$$

where M is a minimal generator of I^q, if and only if n is odd, say $n = 2l + 1$ for some $1 \leq l \leq q$, and

$$M = (x_1 x_2) \cdots (x_{2l-1} x_{2l}) N \text{ with } N \in I^{q-l}.$$

In this case we also have $x_n x_j \in I^{q+1} : \langle M \rangle$ for all $j = 1, \ldots, n$.

Proof Let us start by proving the "if" direction. Suppose that $n = 2l + 1$ and $M = (x_1 x_2) \cdots (x_{2l-1} x_{2l}) N$ with $N \in I^{q-l}$ for some $1 \leq l \leq q$. Then

$$x_n^2 M = (x_n x_1)(x_2 x_3) \cdots (x_{2l} x_n) N \in I^{l+1+q-l} = I^{q+1}.$$

Thus, $x_n^2 \in (I^{q+1} : \langle M \rangle)$.

We proceed to prove the "only if" direction. Indeed, by Theorem 5.27, if $x_n^2 \in (I^{q+1} : \langle M \rangle)$, then x_n must be even-connected to itself with respect to M. Let $x_n = p_0, p_2, \ldots, p_{2l+1} = x_n$ be a shortest even-connected path between x_n and itself.

Consider the case where there exists some $1 \leq j \leq 2l$ such that $p_j = x_n$. If j is odd then $x_n = p_0, \ldots, p_j = x_n$ is a shorter even-connected path between x_n and itself, a contradiction. If j is even, then $x_n = p_j, \ldots, p_{2l+1} = x_n$ is also a shorter even-connected path between x_n and itself, a contradiction. Thus, we may assume that x_n does not appear in the path $p_0, \ldots, p_{2l+1}$ except at its endpoints.

If the path $p_0, \ldots, p_{2l+1}$ is not simple, say for $1 \leq i < j \leq 2l$ we have $p_i = p_j$ (and we choose such i and j so that $j - i$ is minimal), then $p_i, \ldots, p_j$ is a simple closed path lying on C_n. This can only occur if this simple path is in fact C_n, which then violates our assumption about the appearance of x_n in the path $p_0, \ldots, p_{2l+1}$. Therefore, $x_n = p_0, \ldots, p_{2l+1} = x_n$ is a simple closed path on C_n. It follows that $x_n = p_0, \ldots, p_{2l+1} = x_n$ is C_n. This, in particular, implies that $n = 2l + 1$ is odd, and by re-indexing if necessary, we may assume that $p_i = x_i$ for all $i = 1, \ldots, n$. Moreover, by the definition of even-connected path, we have that

$$M = (p_1 p_2) \cdots (p_{2l-1} p_{2l})N = (x_1 x_2) \cdots (x_{2l-1} x_{2l})N,$$

where N is the product of $q - l$ edges in C_n (whence, $N \in I^{q-l}$).

The last statement of the theorem follows from Theorem 5.27 and the following observation: for j odd, $p_0, \ldots, p_j$ is an even-connected path between x_n and x_j; and for j even, $p_j, \ldots, p_{2l+1}$ is an even-connected path between x_j and x_n.

We are now ready to present the proof of Theorem 5.2. Let C_n be a cycle, and let $I = I(C_n)$. The first statement follows from by work of Jacques [118, Theorem 7.6.28]. In fact, all nonzero graded Betti numbers of $I(C_n)$ were computed in [118, Theorems 7.6.28]. Specifically, these numbers are, for $j < n$ and $2i \geq j$,

$$\beta_{i,j}(I(C_n)) = \frac{n}{n - 2(j-l)} \binom{j-i}{2i-j} \binom{n-2(j-i)}{j-i},$$

and

$$\beta_{2m+1,n}(I(C_n)) = 1 \text{ if } n = 3m + 1$$

$$\beta_{2m+1,n}(I(C_n)) = 1 \text{ if } n = 3m + 2$$

$$\beta_{2m,n}(I(C_n)) = 2 \text{ if } n = 3m.$$

We shall prove the second statement of the theorem. In light of Theorem 5.8, it suffices to show that

$$\operatorname{reg} I^q \leq 2q + \nu - 1.$$

where $\nu = \nu(G)$. By applying Theorem 5.25 and using induction, it is enough to prove that

$$\operatorname{reg}(I^{q+1} : \langle M \rangle) \leq \nu + 1 \tag{6.7}$$

for any $s \geq 1$ and any minimal generator M of I^q.

By Theorem 5.27, $I^{q+1} : \langle M \rangle$ is generated in degree 2, and its generators are of the form uv, where either $\{u, v\}$ is an edge in C_n, or u and v are even-connected with respect to M. Observe that if x_n^2 is a generator of $I^{q+1} : \langle M \rangle$, then by Lemma 6.5, we get that n is odd and $x_n x_j$ is a generator of $I^{q+1} : M$ for all $j = 1, \ldots, n$. In this case, in polarizing $I^{q+1} : \langle M \rangle$, we replace the generator x_n^2 by $x_n y_n$, where y_n is a new variable. Thus, if we denote by J the polarization of $I^{q+1} : \langle M \rangle$ then J has the form

$$J = I(G) + \langle x_{i_1} y_{i_1}, \ldots, x_{i_t} y_{i_t} \rangle,$$

where G is a graph over the vertices $\{x_1, \ldots, x_n\}$, $y_{i_1}, \ldots, y_{i_t}$ are new variables, and $x_{i_1}^2, \ldots, x_{i_t}^2$ are all non-squarefree minimal generators of $I^{q+1} : \langle M \rangle$. Note that polarization does not change the regularity, and we have

$$\operatorname{reg} J = \operatorname{reg}(I^{q+1} : \langle M \rangle).$$

Note also that since $I^{q+1} : \langle M \rangle$ have all edges of C_n as minimal generators, G has C_n as a Hamiltonian cycle.

Consider the case that $I^{q+1} : \langle M \rangle$ indeed has non-squarefree minimal generators (i.e., $t \neq 0$). For each $j = 0, \ldots, t$, let H_j be the graph whose edge ideal is $I(G) + \langle x_{i_1} y_{i_1}, \ldots, x_{i_j} y_{i_j} \rangle$. Then, $H_0 = G$ and $J = I(H_t)$.

By Lemma 6.5 (and following our observation above), $\{x_{i_j}, x_l\}$ is an edge in G for any $j = 1, \ldots, t$ and any $l = 1, \ldots, n$. This implies that the induced subgraph $H_j \setminus N_{H_j}[x_{i_j}]$ of H_j consists of isolated vertices $\{y_{i_1}, \ldots, y_{i_{j-1}}\}$. It follows that $\operatorname{reg}(H_j \setminus N_{H_j}[x_{i_j}]) = 0$, and by Remark 5.19, we have

$$\operatorname{reg} H_j = \operatorname{reg}(H_j \setminus x_{i_j}).$$

However, y_{i_j} is an isolated vertex in $H_j \setminus x_{i_j}$ and $H_j \setminus \{x_{i_j}, y_{i_j}\}$ is an induced subgraph of $H_j \setminus y_{i_j} = H_{j-1}$, and so we get

$$\operatorname{reg} H_j = \operatorname{reg}(H_j \setminus x_{i_j}) = \operatorname{reg}(H_j \setminus \{x_{i_j}, y_{i_j}\})$$

$$\leq \operatorname{reg}(H_j \setminus y_{i_j}) = \operatorname{reg}(H_{j-1}). \tag{6.8}$$

Noting that H_{j-1} is an induced subgraph of H_j, and by Lemma 5.18, this implies that $\operatorname{reg} H_{j-1} \leq \operatorname{reg} H_j$. Therefore, coupled with (6.8), we obtain

$$\operatorname{reg} H_j = \operatorname{reg} H_{j-1} \text{ for all } j = 1, \ldots, t.$$

In particular, it follows that

$$\operatorname{reg} J = \operatorname{reg} H_t = \operatorname{reg} H_0 = \operatorname{reg} G.$$

To prove (6.7), it now remains to show that $\operatorname{reg} G \leq \nu + 1$. This follows from Beyarslan, Hà, and Trung [15, Theorems 3.1 and 3.2] by observing that G contains a Hamilton path or Hamilton cycle. The proofs of [15, Theorems 3.1 and 3.2] are inductive on $|V(G)|$ and use (5.3) to remove one vertex at a time. The proof of Theorem 5.2 completes.

Chapter 7
Final Comments and Further Reading

Banerjee's inductive method [7] has also been successfully applied by various authors, such as Alilooee, Beyarslan, and Selvaraja [3], Jayanthan, Narayanan, and Selvaraja [120, 122], and Moghimian, Norouzi Seyed Fakhari, and Yassemi, [139, 144], pushing Theorems 5.1 and 5.2 further to the classes of unicyclic graphs (see Theorem 5.3) and very well-covered graphs (see Theorem 5.4). The core of given arguments in these works is an understanding of ideals of the form $I^{q+1} : \langle M \rangle$, where $I = I(G)$ is the edge ideal of a simple graph G and M is a minimal generator of I^q.

Generally, $I^{q+1} : \langle M \rangle$ may contain squares of variables. By polarization, one can reduce to edge ideals of graphs. The problem becomes to compare the regularity of edge ideals of these resulting graphs to that of G. It is often possible to describe how these resulting graphs are constructed from G, but is difficult to compare the regularities of their edge ideals.

Since the end of our PRAGMATIC school, a number papers have appeared addressing Problem 5.6, Conjecture 5.9, and Questions 5.12 and 5.14 (see [8, 89, 120, 121, 151]). Particularly, Conjecture 5.9 was proved for vertex decomposable graphs by Banerjee, Beyarslan, and Hà [8] and Jayanthan and Selvaraja [121], a weaker version of Conjecture 5.9 was established for all graphs by Seyed Fakhari and Yassemi [151], Question 5.14 was verified by Gu, Hà, O'Rourke, and Skelton [89], and Question 5.12 was also examined for odd cycles by the same group of authors.

Participants in our PRAGMATIC school have also made progress toward problems and questions introduced, and their investigation has resulted in a number of publications. Specifically, Cid-Ruiz in [39] studied the regularity and Gröbner bases of the Rees algebra of edge ideals of bipartite graphs, and Cid-Ruiz, Jafari, Nemati and Picone in [40] examined the regularity of powers of the edge ideal of bicyclic graphs.

Part III
The Containment Problem

Chapter 8
The Containment Problem: Background

The study of ideals underlies both algebra and geometry. For example, the study of homogeneous ideals in polynomial rings is an aspect of both commutative algebra and of algebraic geometry. In both cases, given an ideal, one wants to understand how the ideal behaves. One way in which algebra and geometry differ is in what it means to be "given an ideal". For an algebraist it typically means being given generators of the ideal. For a geometer it often means being given a locus of points (or a scheme) in projective space, the ideal then being all elements of the polynomial ring which vanish on the given locus or scheme. Determining generators for the ideal defining a scheme sometimes requires significant effort, and if given generators a geometer will usually want to know what vanishing locus they cut out. Thus while both algebraists and geometers study ideals, their starting points are different.

These differing starting points lead to other differences. Since for a geometer it is the vanishing locus which counts, one ideal can be swapped for another simpler one with the same vanishing locus. In particular, for the geometer saturated ideals are the main focus of study. This becomes relevant already in the simplest possible geometrical situation, namely finite sets of points in projective space. Given a radical homogeneous ideal defining a finite set of points in projective space, both algebraists and geometers are interested in how powers of the ideal behave. The scheme structure defined on a finite set of points in projective space by a power of the ideal defining the points is an example of what is known as a fat point scheme. But for the geometer what is of most interest is saturations of the powers (which in the case of a radical ideal of points is the same thing as what is known as symbolic powers), since the saturation of a power defines the same scheme as the power. Partly as a way of bridging the gap between algebra and geometry, it has become of interest to study how symbolic powers compare to ordinary powers of ideals and specifically to try to determine which ordinary powers a given symbolic power contains. This problem, which has become known as the containment problem, is the focus of the next section. Here we review background that is useful in studying the containment problem.

E. Carlini et al., *Ideals of Powers and Powers of Ideals*, Lecture Notes of the Unione Matematica Italiana 27, https://doi.org/10.1007/978-3-030-45247-6_8

8.1   Fat Points

Let $p_1, \ldots, p_s \in \mathbb{P}^N$ be distinct points with defining ideals $I(p_1), \ldots, I(p_s) \subseteq \mathbb{K}[\mathbb{P}^n]$. Given integers $m_i \geq 0$, the ideal

$$I = \bigcap_i I(p_i)^{m_i} \subseteq R = \mathbb{K}[\mathbb{P}^N] = \mathbb{K}[x_0, \ldots, x_N]$$

is homogeneous, meaning $f \in I$ if and only if every homogeneous component f_t of f is in I. Thus I is the direct sum $I = \bigoplus_t [I]_t$, where $[I]_t$ is the vector space span of the homogeneous elements in I of degree t.

Definition 8.1 The ideal I defines a 0-dimensional subscheme denoted $Z = m_1 p_1 + \cdots + m_s p_s \subseteq \mathbb{P}^N$ which is called a *fat point* subscheme.

We denote the ideal I by $I = I(Z)$. When we refer to the *degree* $\deg(Z)$ of Z, we will mean the scheme theoretic degree, hence $\deg(Z) = \sum_i \binom{m_i + N - 1}{N}$.

8.2   Blow Ups and Sheaf Cohomology

We refer to Hartshorne [103] for general background on divisors, their associated line bundles and sheaf cohomology. However, the next fact will often allow us to avoid dealing with some of this background. Let $\pi : X \to \mathbb{P}^N$ be the blow up of the points p_i with L being the pullback to X of a general hyperplane and let $E_i = \pi^{-1}(p_i)$. The group $\mathrm{Cl}(X)$ of linear equivalence classes of divisors on X is a free abelian group with basis given by the divisor classes $[L], [E_1], \ldots, [E_s]$. In case $N = 2$, this is an orthogonal basis for the intersection form on $\mathrm{Cl}(X)$, where $-L^2 = E_1^2 = \cdots = E_s^2 = -1$. The canonical class of X, denoted K_X, is an important divisor class; with respect to this basis it is $-K_X = 3[L] - [E_1] - \cdots - [E_s]$. We will sometimes abuse notation and write $-K_X = 3L - E_1 - \cdots - E_s$. Given $Z = m_1 p_1 + \cdots + m_s p_s$, it is convenient to denote $m_1 E_1 + \cdots + m_s E_s$ by E_Z. Thus for $Z = p_1 + \cdots + p_s$ we have $-K_X = 3L - E_Z$.

Theorem 8.2 ([97, Proposition IV.1.1]) *There is a canonical $\mathbb{K}$-vector space isomorphism*

$$H^0(X, \mathscr{O}_X(tL - E_Z)) \cong [I(Z)]_t.$$

8.3   Hilbert Functions

Let $N \geq 1$ and let $I \subseteq R = \mathbb{K}[\mathbb{P}^N]$ be a homogeneous ideal, for example $I = I(Z)$ for a fat point scheme $Z = m_1 p_1 + \cdots + m_s p_s$.

Definition 8.3 The *Hilbert function* of I is the function $h(t, I) = \dim_{\mathbb{K}}[I]_t$ giving the vector space dimension of $[I]_t$.

We define $\alpha(I)$ to be the least degree t such that $h(t, I) > 0$. This is defined as long as $I \neq (0)$. Note that $h(t, I)$ is strictly increasing for $t \geq \alpha(I)$.

Given a fat point scheme Z, it is also convenient to define $h_Z(t) = \dim_{\mathbb{K}} R_t - h(t, I(Z))$, which we call the *Hilbert function of* Z; it is the Hilbert function of $R/I(Z)$ since $R/I(Z)$ is graded and we have

$$\dim_{\mathbb{K}}[R/I(Z)]_t = \dim_{\mathbb{K}} R_t - \dim_{\mathbb{K}} I(Z)_t = h_Z(t).$$

It can be shown that $h_Z(t) = \deg Z$ for $t \gg 0$ (see Geramita and Maroscia [81, Proposition 1.1]; note that to apply this result, you need to observe that $R/I(Z)$ is a Cohen-Macaulay ring of dimension 1.)

Define $\Delta h_Z(0) = 1$ and $\Delta h_Z(t) = h_Z(t) - h_Z(t-1)$ for $t > 0$. Then it is known that Δh_Z is unimodal (i.e., it starts out nondecreasing, then becomes nonincreasing), nonnegative and has $\Delta h_Z(t) = 0$ for $t \gg 0$ (again, see [81, Proposition 1.1]). We define the *regularity* $\mathrm{reg}(I(Z))$ of $I(Z)$ to be the least t such that $\Delta h_Z(t) = 0$. Note that one can show that when $I = I(Z)$, this definition of regularity agrees with definition of regularity given in Definition 4.4.

8.4 Waldschmidt Constants: Asymptotic α

Given points $p_i \in \mathbb{P}^N$, let $Z = m_1 p_1 + \cdots + m_s p_s \subseteq \mathbb{P}^N$. Its ideal is

$$I(Z) = I(p_1)^{m_1} \cap \cdots \cap I(p_s)^{m_s},$$

and the *m*th *symbolic power* of I, denoted $I^{(m)}$, is the saturation of $(I(Z))^m$, which can be shown to be $I(Z)^{(m)} = I(mZ) = I(p_1)^{mm_1} \cap \cdots \cap I(p_s)^{mm_s}$. One can define symbolic powers more generally, but doing so involves technicalities which we avoid for now.

It is not hard to show that $(I(Z))^r \subseteq (I(Z))^{(r)}$ and hence that $\alpha((I(Z))^{(r)}) \leq \alpha((I(Z))^r)$. And if $I, J \subseteq \mathbb{K}[\mathbb{P}^N]$ are nonzero homogeneous ideals, then $\alpha(IJ) = \alpha(I) + \alpha(J)$. In particular, we have $\alpha((I(Z))^r) = r\alpha(I(Z))$ and thus $\alpha((I(Z))^{(r)}) \leq r\alpha(I(Z))$. Easy examples show that $\alpha((I(Z))^{(r)}) < r\alpha(I(Z))$ can occur (for example $\alpha((I(Z))^{(2)}) = 3 < 4 = 2\alpha(I(Z))$ for $Z = p_1 + p_2 + p_3 \subset \mathbb{P}^2$ with p_1, p_2, p_3 noncollinear).

We now define an asymptotic version of α.

Definition 8.4 Let $Z = m_1 p_1 + \cdots + m_s p_s$ be a nonzero fat point subscheme of $\mathbb{P}^N$. The *Waldschmidt constant* $\widehat{\alpha}(I(Z))$ of $I(Z)$ is

$$\widehat{\alpha}(I(Z)) = \inf\left\{\frac{\alpha((I(Z))^{(m)})}{m} \,\middle|\, m > 0\right\}.$$

Many of the properties of the Waldschmidt constant of fat point subschemes can be found in Harbourne's paper [98]. We record these properties as exercises.

Exercise 8.5 (See [98, Example 1.3.3]) Let X be the surface obtained by blowing up distinct points $p_1, \ldots, p_r$. Let $I(Z)$ be the ideal of $Z = m_1 p_1 + \cdots + m_r p_r$ and let $F_{t,m} = tL - mE_Z$. Then

$$\widehat{\alpha}(I(Z)) = \inf\left\{ \frac{t}{m} \,\middle|\, h^0(X, \mathscr{O}_X(F_{t,m})) > 0 \right\}.$$

In fact $\widehat{\alpha}(I(Z))$ is a limit, as seen in the exercise below.

Exercise 8.6 (See [98, Example 1.3.4])   Let Z be a nonzero fat point subscheme of $\mathbb{P}^N$.

(a) Then $1 \le \widehat{\alpha}(I(Z)) \le \sum_i m_i$.
(b) Let m, n be positive integers. Then

$$\alpha(I((m+n)Z)) \le \alpha(I(mZ)) + \alpha(I(nZ)).$$

(c) Let m, n be positive integers. Then

$$\frac{\alpha(I(mnZ))}{mn} \le \frac{\alpha(I(mZ))}{m}.$$

(d) Fekete's Subadditivity Lemma [73] implies for each n that

$$\widehat{\alpha}(I(Z)) = \lim_{m \to \infty} \frac{\alpha(I(mZ))}{m} \le \frac{\alpha(I(nZ))}{n}.$$

(e) We have $\widehat{\alpha}(I(nZ)) = n\,\widehat{\alpha}(I(Z))$.
(f) Over the complexes, Waldschmidt and Skoda [154, 167] obtained the bound

$$\frac{\alpha(I(Z))}{N} \le \widehat{\alpha}(I(Z))$$

using some rather hard analysis. A proof using multiplier ideals is given in [131]. Here is another approach which everyone now takes for granted but which in fact was first used by Harbourne and Roé [100, p. 2] and first appears explicitly in Harbourne and Huneke [99]. It is known that

$$I((N + m - 1)rZ) \subseteq I(mZ)^r$$

for all $m, r > 0$ by work of Ein-Lazarsfeld-Smith and Hochster-Huneke [62, 115]. Assuming this, one can show for each $n > 0$ that

$$\frac{\alpha(I(mZ))}{N + m - 1} \le \widehat{\alpha}(I(Z))$$

and hence that

$$\frac{\alpha(I(mZ))}{N+m-1} \leq \widehat{\alpha}(I(Z)) \leq \frac{\alpha(I(mZ))}{m}.$$

Remark 8.7 Consider $Z = p_1 + \cdots + p_s$ for distinct points $p_i \in \mathbb{P}^N$. When $N = 2$ Chudnovsky [38] gave the following slightly improved lower bound, based on a clever algebraic-geometric idea that his paper does not write out explicitly (see the end of section 3 of [98] for this algebraic-geometric proof; for a more algebraic proof, see [99]):

$$\frac{\alpha(Z)+1}{2} \leq \widehat{\alpha}(I(Z)).$$

For $N > 2$, Chudnovsky conjectured that

$$\frac{\alpha(Z)+N-1}{N} \leq \widehat{\alpha}(I(Z))$$

should hold, but this is still open in general. See [133] for additional discussion, including a generalized version of the conjecture due to Demailly, and see Esnault and Viehweg [69] for a partial result.

One can give a universal upper bound for $\widehat{\alpha}(I(Z))$ depending only on the m_i and N. When this bound is not rational it is an open problem to show whether it is ever attained.

Exercise 8.8 (See [98, Example 1.3.7].) Let $Z = m_1 p_1 + \cdots + m_s p_s$ be a nonzero fat point subscheme of $\mathbb{P}^N$. Then $\widehat{\alpha}(I(Z)) \leq \sqrt[N]{\sum_i m_i^N}$.

Remark 8.9 When $N = 2, s > 9, m_i = 1$ for all i and the points p_i are sufficiently general (such as generic points $p_i = [a_i : b_i : c_i], 1 \leq i \leq s$, meaning the ratios $a_1/c_1, b_1/c_1, \ldots, a_s/c_s, b_s/c_s$ are algebraically independent over the prime field), a famous conjecture of Nagata [141] is equivalent to the bound in Exercise 8.8 being an equality. As a step in his counterexample to Hilbert's 14th Problem, he verified this in the case s is a square. It is otherwise still open.

It is not in general known how to determine $\widehat{\alpha}(I(Z))$ exactly. However, by Exercise 8.6(f), one can compute $\widehat{\alpha}(I(Z))$ arbitrarily accurately by computing $\alpha(I(mZ))$ for large m. Thus for any real number $a \neq \widehat{\alpha}(I(Z))$, it is just a computation to show that $a \neq \widehat{\alpha}(I(Z))$. For example, we have the following (see [98, Corollary 1.3.8] for the easy proof).

Corollary 8.10 Let $Z = m_1 p_1 + \cdots + m_s p_s \subseteq \mathbb{P}^N$ be a nonzero fat point subscheme. Let t be rational. If $t > \widehat{\alpha}(I(Z))$, then $\dim_{\mathbb{K}}[I(mZ)]_{mt} > 0$ for all $m \gg 0$ such that mt is an integer, and if $t < \widehat{\alpha}(I(Z))$, then $\dim_{\mathbb{K}}[I(mZ)]_{mt} = 0$ for all $m > 0$ such that mt is an integer.

Chapter 9
The Containment Problem

Given a fat point scheme $Z = m_1 p_1 + \cdots + m_s p_s \subset \mathbb{P}^N$, the containment problem for Z is to determine for which r and m the containment $(I(Z))^{(m)} \subseteq (I(Z))^r$ holds. In this section we present some initial results for the containment problem, and we define an asymptotic quantity, the resurgence, that measure to what extent the containment hold for a given Z.

9.1 Containment Problems

Let $Z = m_1 p_1 + \cdots + m_s p_s \subseteq \mathbb{P}^N$ be a fat point subscheme. As noted in the previous section, $I(rZ)$ is the saturation of $(I(Z))^r$. Thus $I(Z)^r = Q \cap I(rZ)$ for some M-primary ideal Q, where $M = (x_0, \ldots, x_N)$ (and hence Q contains a power of M, from which it follows that $[Q]_t = [M]_t$ for all $t \gg 0$). In particular, we have that $I(Z)^m \subseteq I(Z)^{(m)}$ for all $m \geq 1$ and $[I(Z)^r]_t = [I(rZ)]_t$ for all $t \gg 0$.

Exercise 9.1 (See [98, Example 3.1.1]) Let $I = I(Z)$ for $Z = m_1 p_1 + \cdots + m_s p_s \subseteq \mathbb{P}^N$ with $m_i > 0$ for all i. Then:

(a) $I^m \subseteq I^r$ if and only if $m \geq r$.
(b) $I^{(m)} \subseteq I^{(r)}$ if and only if $m \geq r$.
(c) $I^m \subseteq I^{(r)}$ if and only if $m \geq r$.
(d) $I^{(m)} \subseteq I^r$ implies $m \geq r$, but $m \geq r$ does not in general imply $I^{(m)} \subseteq I^r$.

It is not known in general how to solve the containment problem for a given Z, but it is useful to know that for $m \gg 0$ we always do have containment. We first introduce the following term.

Definition 9.2 The *saturation degree* of I^r, denoted $\mathrm{satdeg}(I^r)$, is the least t such that $(I^r)_j = (I^{(r)})_j$ for all $j \geq t$.

E. Carlini et al., *Ideals of Powers and Powers of Ideals*, Lecture Notes of the Unione Matematica Italiana 27, https://doi.org/10.1007/978-3-030-45247-6_9

The following result is [98, Proposition 3.1.2].

Proposition 9.3 *Let $I = I(Z)$ be a fat point scheme $Z \subseteq \mathbb{P}^N$. If*

$$m \geq \max\left\{\frac{\text{satdeg}(I^r)}{\widehat{\alpha}(I(Z))}, r\right\},$$

then $I^{(m)} \subseteq I^r$.

Proof Since $m \geq r$, we have $I^{(m)} \subseteq I^{(r)}$. Since $m \geq \text{satdeg}(I^r)$, if $[I^{(m)}]_t \neq 0$, then $t \geq \alpha(I(mZ)) \geq m\widehat{\alpha}(I(Z)) \geq \text{satdeg}(I^r)$, so $[I^{(m)}]_t \subseteq [I^{(r)}]_t = [I^r]_t$. Hence $I^{(m)} \subseteq I^r$.

It turns out there is a very nice universal bound (given by Ein, Lazarsfeld, and Smith [62] and Hochster and Huneke [115], motivated by work of Swanson [158]) depending only on N and r for how big m must be to ensure that the containment $I^{(m)} \subseteq I^r$ holds. We give a version of this result in Theorem 9.4. (See Definition 10.1 for how to define symbolic powers for ideals that need not be ideals of fat points. This is the definition used in Hochster-Huneke [115]. It has the property that $I^{(1)} = I$, and that $I^{(m)} = I^m$ for all $m \geq 1$ when I is not saturated.)

Theorem 9.4 *Let $I \subseteq \mathbb{K}[\mathbb{P}^N]$ be a homogeneous ideal, and let $r, s \geq 1$. Then we have $I^{(r(s+N-1))} \subseteq (I^{(s)})^r$. In particular (taking $s = 1$), if $m \geq rN$, then we have $I^{(m)} \subseteq I^r$ (since $I^{(m)} \subseteq I^{(rN)} \subseteq I^r$).*

With this result in hand, one can ask: can one do better? There are various ways to think about what being better means here. A result of Bocci and Harbourne [21] shows that no constant less than N suffices:

Theorem 9.5 *If $c < N$, then there is an $r > 0$ and $m > cr$ such that $I^{(m)} \not\subseteq I^r$ for some $I = I(Z)$, where $Z = p_1 + \cdots + p_s \subseteq \mathbb{P}^N$ for distinct points p_i.*

There is another way to think about this which has led to a fair amount of recent research under the rubric of Harbourne-Huneke bounds (see for example Walker [168]). This point of view began with a question of C. Huneke: if $I = I(Z)$ is a radical ideal defining a finite set of points $Z \subset \mathbb{P}^2$, is it always true that $(I(Z))^{(3)} \subseteq (I(Z))^2$? In thinking about this, Harbourne was led to make the following conjecture (see [9, Conjecture 8.4.2]).

Conjecture 9.6 Let $Z \subseteq \mathbb{P}^N$ be a fat point subscheme. Then

$$I((Nr - N + 1)Z) \subseteq I(Z)^r$$

holds for each $r > 1$.

It turns out the containment $I((Nr - N + 1)Z) \subseteq I(Z)^r$ often holds, but not always. The first failure of containment was given by Dumnicki, Szemberg, and Tutaj-Gasińska [58], for which the ground field was the complex numbers and $N =$

$r = 2$. Additional failures are given by Dumnicki et al. [59], but so far no failures over the complex numbers are known if either $r > 2$ or $N > 2$. This suggests that perhaps the following problem at least has an affirmative answer, but this problem remains open.

Question 9.7 Let $Z \subseteq \mathbb{P}^N$ be a fat point subscheme. Then is it always true that

$$I((Nr - N + 1)Z) \subseteq I(Z)^r$$

for all $r \gg 0$?

Here is a useful criterion for containment to fail.

Exercise 9.8 (See [98, Example 3.1.7]) Let $Z \subseteq \mathbb{P}^N$ be a fat point subscheme, $I = I(Z)$. If $\alpha(I^{(m)}) < r\alpha(I)$, then $I^{(m)} \not\subseteq I^r$.

9.2 The Resurgence

As we saw, for a fat point subscheme $Z \subset \mathbb{P}^N$, the containment $(I(Z))^{(m)} \subseteq (I(Z))^r$ is guaranteed for $m \geq Nr$ for all Z, but how small can m be for a specific Z? This question led to the definition of the resurgence by Bocci and Harbourne [21].

Definition 9.9 Given a fat point scheme $Z \subseteq \mathbb{P}^N$, define the *resurgence* $\rho(I)$ for $I = I(Z)$ to be

$$\rho(I(Z)) = \sup \left\{ \frac{m}{r} \;\middle|\; I^{(m)} \not\subseteq I^r \right\}.$$

(See Guardo, Harbourne, and Van Tuyl [91] for an asymptotic version of the resurgence.)

Given the ideal I of a fat point subscheme of projective space, the following facts are useful when studying resurgences (see [21]). First, by definition of regularity (Definition 4.1), it follows that the homogeneous generators of I have degree at most $\text{reg}(I)$. Moreover, $[I^r]_t = [I^{(r)}]_t$ holds for $t \geq r \,\text{reg}(I)$ and even for $t \geq \text{reg}(I^r)$ (because $r \,\text{reg}(I) \geq \text{reg}(I^r) \geq \text{satdeg}(I^r)$; see [21]).

The following result is from [21]; we follow the exposition given in [98, Theorem 3.2.4].

Theorem 9.10 *Let $I = I(Z)$ for a nonempty fat point subscheme $Z \subseteq \mathbb{P}^N$.*

(a) We have $1 \leq \rho(I) \leq N$.
(b) If $m/r < \frac{\alpha(I)}{\hat{\alpha}(I)}$, then for all $t \gg 0$ we have $I^{(mt)} \not\subseteq I^{rt}$.
(c) If $m/r \geq \frac{\text{reg}(I)}{\hat{\alpha}(I)}$, then $I^{(m)} \subseteq I^r$.

(d) We have

$$\frac{\alpha(I)}{\widehat{\alpha}(I)} \le \rho(I) \le \frac{\mathrm{reg}(I)}{\widehat{\alpha}(I)},$$

hence $\frac{\alpha(I)}{\widehat{\alpha}(I)} = \rho(I)$ if $\alpha(I) = \mathrm{reg}(I)$.

Proof

(a) By Theorem 9.4, we have $\rho(I) \le N$. By Exercise 9.1(d), we have $\rho(I) \ge 1$.

(b) If $m/r < \frac{\alpha(I)}{\widehat{\alpha}(I)}$, then $\widehat{\alpha}(I) < r\alpha(I)/m$, so for $t \gg 0$ we have

$$\widehat{\alpha}(I) \le \alpha(I^{(mt)})/(mt) < r\alpha(I)/m = rt\alpha(I)/(mt) = \alpha(I^{rt})/(mt),$$

so also $\alpha(I^{(mt)}) < \alpha(I^{rt})$, hence $I^{(mt)} \not\subseteq I^{rt}$ by Exercise 9.8.

(c) Now say $m/r \ge \frac{\mathrm{reg}(I)}{\widehat{\alpha}(I)}$. Then $\alpha(I^{(m)}) \ge m\widehat{\alpha}(I) \ge r\,\mathrm{reg}(I)$. If $t < \alpha(I^{(m)})$, then $[I^{(m)}]_t = (0) \subseteq I^r$. If $t \ge \alpha(I^{(m)})$, then $t \ge r\,\mathrm{reg}(I)$ hence $[I^{(m)}]_t \subseteq [I^{(r)}]_t = [I^r]_t$. Thus $I^{(m)} \subseteq I^r$.

(d) This follows from (b) and (c).

Problem 9.7 has a negative answer if and only if there is an $I = I(Z)$ with $\rho(I) = N$, but no such Z is currently known. There are however a lot of examples with $\rho(I) = 1$. For example, if $I^{(m)} = I^m$ for all $m > 0$, then $\rho(I) = 1$, but it is not known if the converse is true. As the next theorem shows, there is a nice family of ideals that satisfy $I^{(m)} = I^m$.

Theorem 9.11 ([172, Lemma 5, Appendix 6]) *Let I be a homogeneous ideal generated by a regular sequence in $R = \mathbb{K}[x_0, \ldots, x_n]$, i.e., I is a complete intersection. Then*

$$I^{(m)} = I^m \quad \text{for all } m \ge 1.$$

Chapter 10
The Waldschmidt Constant of Squarefree Monomial Ideals

The last two chapters introduced the Waldschmidt constant of a homogeneous ideal of set of (fat) points and some of its properties. In fact, the definition of the Waldschmidt constant makes sense for any homogeneous ideal. In this chapter we explain how to compute this invariant in the case of squarefree monomial ideals. In the case of edge ideals, we will also give a combinatorial interpretation of this invariant. Throughout this chapter, $R = \mathbb{K}[x_1, \ldots, x_n]$ is a polynomial ring over a field $\mathbb{K}$, where $\mathbb{K}$ has characteristic zero and is algebraically closed. All ideals $I \subseteq R$ will be assumed to be homogeneous, and in most cases, I will be a squarefree monomial ideal.

10.1 The Waldschmidt Constant (General Case)

In the previous two chapters, we defined the Waldschmidt constant for ideals of fat points. However, the definition extends quite naturally to any homogeneous radical ideal.

We set

$$\alpha(I) = \min\{i \mid \text{exist } 0 \neq F \in I \text{ with } \deg F = i\}.$$

That is, $\alpha(I)$ is the smallest degree of a minimal generator of I. We recall the definition of a symbolic power of an ideal.

Definition 10.1 Let I be a homogeneous ideal in R. Then the *m-th symbolic power* of I is the ideal

$$I^{(m)} = \bigcap_{P \in \text{ass}(I)} (I^m R_P \cap R)$$

where $I^m R_P$ denotes the ideal of I^m in the ring R_P, i.e., the ring R localized at P.

E. Carlini et al., *Ideals of Powers and Powers of Ideals*, Lecture Notes of the Unione Matematica Italiana 27, https://doi.org/10.1007/978-3-030-45247-6_10

We now recall the invariant of interest from Chap. 8:

Definition 10.2 Let I be a homogeneous radical ideal of R. The *Waldschmidt constant* of I is

$$\widehat{\alpha}(I) := \lim_{m \to \infty} \frac{\alpha(I^{(m)})}{m}.$$

Remark 10.3 In Chap. 8, the Waldschmidt constant was defined in Definition 8.4 as the infimun of the values $\frac{\alpha(I^{(m)})}{m}$ over all integers $m > 0$. The fact that these two definitions are equal can be found in [21, Lemma 2.3.1].

10.2   The Squarefree Monomial Case

In general, computing the Waldschmidt constant of an ideal is quite difficult. Indeed, both Chudnosky's Conjecture ([38], or see Remark 8.7) and Nagata's Conjecture ([141], or see Conjecture 8.9) can be restated as conjectures about the Waldschmidt constant of the ideal of a set of points. However, when I is a squarefree monomial ideal, Bocci, Cooper, Guardo, Harbourne, Janssen, Nagel, Seceleanu, Van Tuyl, and Vu showed in [22] that there is a procedure to compute $\widehat{\alpha}(I)$. We will describe this procedure.

Recall that we call a monomial ideal I a squarefree monomial ideal if it is generated by squarefree monomials. That is, each generator of I has the form $m = x_1^{a_1} \cdots x_n^{a_n}$ with $a_i \in \{0, 1\}$ for all i. The following theorem summarizes some of the nice features of squarefree monomial ideals.

Theorem 10.4 *Let I be a squarefree monomial ideal in $R = \mathbb{K}[x_1, \ldots, x_n]$.*

(i) *There exist unique prime ideals of the form $P_i = \langle x_{i,1}, \ldots, x_{i,t_i} \rangle$ such that $I = P_1 \cap \cdots \cap P_s$.*

(ii) *With the P_i's as above, the m-th symbolic power of I is given by $I^{(m)} = P_1^m \cap \cdots \cap P_s^m$.*

(iii) *For all integers $m \geq 1$,*

$$\alpha(I^{(m)}) = \min\{a_1 + \cdots + a_n \mid x_1^{a_1} \cdots x_n^{a_n} \in I^{(m)}\}.$$

Proof For (i), first note that a squarefree monomial ideal I is radical, that is, $I = \sqrt{I}$. Consequently, $I = P_1 \cap \cdots \cap P_s$ where the P_is run over all the minimal associated prime ideals of I. Furthermore, since these associated prime ideals are all minimal, they are unique. Finally, by Lemma 2.4, each $P_i = \langle x_{i,1}, \ldots, x_{i,t_i} \rangle$. For more details, see Chapter 1 of Herzog and Hibi's book [106], and in particular, Lemma 1.5.4. Statement (ii) can be found in [106, Proposition 1.4.4]; note this result can also be viewed as a special case of work of Cooper, Embree, Hà, and

Hoefel [45, Theorem 3.7]. Statement (iii) follows directly from the definition of $\alpha(-)$ and a monomial ideal.

Example 10.5 We consider the following squarefree monomial ideal which we will use as our running example. In particular, we look at

$$I = \langle x_1 x_2, x_2 x_3, x_3 x_4, x_4 x_5, x_5 x_1 \rangle \subseteq \mathbb{K}[x_1, \ldots, x_5].$$

This ideal has the following primary decomposition

$$I = \langle x_1, x_3, x_4 \rangle \cap \langle x_2, x_4, x_5 \rangle \cap \langle x_3, x_5, x_1 \rangle \cap \langle x_4, x_1, x_2 \rangle \cap \langle x_5, x_2, x_3 \rangle.$$

The next result enables us to determine if a particular monomial belongs to $I^{(m)}$.

Lemma 10.6 *Let $I \subseteq R$ be a squarefree monomial ideal with minimal primary decomposition $I = P_1 \cap P_2 \cap \cdots \cap P_s$ with $P_i = \langle x_{i,1}, \ldots, x_{i,t_i} \rangle$ for $i = 1, \ldots, s$. Then $x_1^{a_1} \cdots x_n^{a_n} \in I^{(m)}$ if and only if $a_{i,1} + \cdots + a_{i,t_i} \geq m$ for $i = 1, \ldots, s$.*

Proof By Theorem 10.4 (ii), $I^{(m)} = P_1^m \cap \cdots \cap P_s^m$. So $x_1^{a_1} \cdots x_n^{a_n} \in I^{(m)}$ if and only if $x_1^{a_1} \cdots x_n^{a_n}$ is in P_j^m for all $j = 1, \ldots, s$. This happens if and only if there exists at least one generator $f_j \in P_j^m$ such that f_j divides $x_1^{a_1} \cdots x_n^{a_n}$ (for $j = 1, \ldots, s$), which is equivalent to requiring $a_{j_1} + \cdots + a_{j_{s_j}} \geq m$ for $j = 1, \ldots, s$.   ∎

The above lemma is the key observation that is needed in order to determine the Waldschmidt constant of squarefree monomial ideals. To make this more precise, let's return to our running example.

Example 10.7 Let I be as in Example 10.5. To determine if $x_1^{a_1} x_2^{a_2} x_3^{a_3} x_4^{a_4} x_5^{a_5} \in I^{(m)}$, Lemma 10.6 says we need to find integers $a_1, \ldots, a_5$ that satisfy the following inequalities

$$a_1 + a_3 + a_4 \geq m$$
$$a_2 + a_4 + a_5 \geq m$$
$$a_3 + a_5 + a_1 \geq m$$
$$a_4 + a_1 + a_2 \geq m$$
$$a_5 + a_2 + a_3 \geq m.$$

If we also want to find $\alpha(I^{(m)})$, we also need to find the tuple $(a_1, a_2, a_3, a_4, a_5)$ that not only satisfies the above inequalities, but also minimizes $a_1 + a_2 + a_3 + a_4 + a_5$ (by Theorem 10.4).

Stepping back for a moment, notice in the above example, we have described the computation of $\alpha(I^{(m)})$ as a solution to a linear program. This idea can be extended to all squarefree monomial ideals.

In particular, given a primary decomposition of our squarefree monomial ideal $I = P_1 \cap \cdots \cap P_s$, we define an $s \times n$ matrix A where

$$A_{i,j} = \begin{cases} 1 \text{ if } x_j \in P_i \\ 0 \text{ if } x_j \notin P_i. \end{cases}$$

We then define our linear program (LP) constructed from I as follows:

minimize $\mathbf{1}^T \mathbf{y}$

subject to $A\mathbf{y} \geq \mathbf{1}$ *and* $\mathbf{y} \geq \mathbf{0}$.

Here $\mathbf{y}^T = \begin{bmatrix} y_1 & \cdots & y_n \end{bmatrix}$, and $\mathbf{1}$, respectively $\mathbf{0}$, is an appropriate sized vector of 1's, respectively 0's.

Example 10.8 Continuing with our running example, the ideal I of Example 10.5 gives us the following LP:

$$\text{minimize } \mathbf{1}^T \mathbf{y} = y_1 + y_2 + y_3 + y_4 + y_5$$

$$\text{subject to } \begin{bmatrix} 1 & 0 & 1 & 1 & 0 \\ 0 & 1 & 0 & 1 & 1 \\ 1 & 0 & 1 & 0 & 1 \\ 1 & 1 & 0 & 1 & 0 \\ 0 & 1 & 1 & 0 & 1 \end{bmatrix} \begin{bmatrix} y_1 \\ y_2 \\ y_3 \\ y_4 \\ y_5 \end{bmatrix} \geq \begin{bmatrix} 1 \\ 1 \\ 1 \\ 1 \\ 1 \end{bmatrix} \text{ and } \begin{bmatrix} y_1 \\ y_2 \\ y_3 \\ y_4 \\ y_5 \end{bmatrix} \geq \begin{bmatrix} 0 \\ 0 \\ 0 \\ 0 \\ 0 \end{bmatrix}.$$

We now come to our theorem which relates this idea of a LP to the Waldschmidt constant.

Theorem 10.9 ([22, Theorem 3.2]) *Let $I \subseteq R$ be a squarefree monomial ideal with minimal primary decomposition $I = P_1 \cap P_2 \cap \cdots \cap P_s$ with $P_i = \langle x_{i,1}, \ldots, x_{i,t_i} \rangle$ for $i = 1, \ldots, s$. Let A be the $s \times n$ matrix where*

$$A_{i,j} = \begin{cases} 1 & \text{if } x_j \in P_i \\ 0 & \text{if } x_j \notin P_i. \end{cases}$$

Consider the following LP:

minimize $\mathbf{1}^T \mathbf{y}$

subject to $A\mathbf{y} \geq \mathbf{1}$ *and* $\mathbf{y} \geq \mathbf{0}$

and suppose that $\mathbf{y}^*$ is a feasible solution (i.e., $\mathbf{y}^*$ is a vector that satisfies the LP) that realizes the optimal value. Then

$$\widehat{\alpha}(I) = \mathbf{1}^T \mathbf{y}^*.$$

That is, $\widehat{\alpha}(I)$ is the optimal value of the LP.

Proof Let $(\mathbf{y}^*)^T = \begin{bmatrix} y_1^* & y_2^* & \cdots & y_n^* \end{bmatrix}$ be a feasible solution to the LP that also realizes the optimal solution. Because the matrix A in the LP contains only integer entries, all the y_i's are rational (see [149, Corollary 1.3.1]; note that this result applies to our situation since the equations in our LP all have integer coefficients). So, we can write $(\mathbf{y}^*)^T = \begin{bmatrix} \frac{a_1}{b_1} & \frac{a_2}{b_2} & \cdots & \frac{a_n}{b_n} \end{bmatrix}$ with integers a_i, b_i for $i = 1, \ldots, n$.

Set $b = \mathrm{lcm}(b_1, \ldots, b_n)$. Then $A(b\mathbf{y}) \geq \mathbf{b}$ where $\mathbf{b}$ is an s-vector of b's. So, $(b\mathbf{y})$ is a feasible integer solution to the system $A\mathbf{z} \geq \mathbf{b}$. In other words, for each $j = 1, \ldots, s$,

$$b\left(\frac{a_{j_1}}{b_{j_1}} + \cdots + \frac{a_{j_{s_j}}}{b_{j_{s_j}}}\right) = \frac{ba_{j_1}}{b_{j_1}} + \cdots + \frac{ba_{j_{s_j}}}{b_{j_{s_j}}} \geq b.$$

It then follows by Lemma 10.6 that

$$x_1^{\frac{ba_1}{b_1}} x_2^{\frac{ba_2}{b_2}} \cdots x_n^{\frac{ba_n}{b_n}} \in I^{(b)}.$$

Thus,

$$\alpha(I^{(b)}) \leq \frac{ba_1}{b_1} + \frac{ba_2}{b_2} + \cdots + \frac{ba_n}{b_n},$$

or equivalently,

$$\widehat{\alpha}(I) \leq \frac{\alpha(I^{(b)})}{b} \leq \frac{a_1}{b_1} + \frac{a_2}{b_2} + \cdots + \frac{a_n}{b_n} = \mathbf{1}^T\mathbf{y}^*$$

since $\widehat{\alpha}(I) \leq \frac{\alpha(I^{(b)})}{b}$ for all $b > 0$ by Remark 10.3.

To show the reverse inequality, suppose for a contradiction that $\widehat{\alpha}(I) < \mathbf{1}^T\mathbf{y}^*$. Since $\widehat{\alpha}(I) = \inf\left\{\alpha(I^{(m)})/m\right\}_{m \in \mathbb{N}}$, there must exist some m such that

$$\frac{\alpha(I^{(m)})}{m} < \frac{a_1}{b_1} + \frac{a_2}{b_2} + \cdots + \frac{a_n}{b_n} = \mathbf{1}^T\mathbf{y}^*.$$

Let $x_1^{e_1} x_2^{e_2} \cdots x_n^{e_n} \in I^{(m)}$ be a monomial with $e_1 + \cdots + e_n = \alpha(I^{(m)})$. Then, by Lemma 10.6, we have

$$e_{j_1} + \cdots + e_{j_{s_j}} \geq m \quad \text{for all } j = 1, \ldots, s.$$

In particular, if we divide all the s equations by m, we have

$$\frac{e_{j_1}}{m} + \cdots + \frac{e_{j_{s_j}}}{m} \geq 1 \quad \text{for all } j = 1, \ldots, s.$$

But then $\mathbf{w}^T = \begin{bmatrix} \frac{e_1}{m} & \cdots & \frac{e_s}{m} \end{bmatrix}^T$ satisfies $A\mathbf{w} \geq \mathbf{1}$ and $\mathbf{w} \geq \mathbf{0}$. That is, $\mathbf{w}$ is a feasible solution to the LP, and furthermore, $\frac{\alpha(I^{(m)})}{m} = \mathbf{1}^T \mathbf{w} < \frac{a_1}{b_1} + \frac{a_2}{b_2} + \cdots + \frac{a_n}{b_n} = \mathbf{1}^T \mathbf{y}^*$, contradicting the fact that $\mathbf{1}^T \mathbf{y}^*$ is the optimal value of the LP.

Remark 10.10 Because the Waldschmidt constant of a squarefree monomial ideal can be formulated in terms of a LP, it can be solved by using the *simplex method* developed by Dantzig in the 1940s. There are a number of online calculators that will allow you to solve a LP. Here is one example: http://comnuan.com/cmnn03/cmnn03004/.

Remark 10.11 Note that to set up the LP to find the Waldschmidt constant of a squarefree monomial ideal, we only need to know information about the primary decomposition of the monomial ideal I.

Example 10.12 For the LP in Example 10.8, the feasible solution that gives the optimal value is

$$\mathbf{y}^T = \begin{bmatrix} \frac{1}{3} & \frac{1}{3} & \frac{1}{3} & \frac{1}{3} & \frac{1}{3} \end{bmatrix}.$$

Consequently,

$$\widehat{\alpha}(I) = \frac{1}{3} + \cdots + \frac{1}{3} = \frac{5}{3}.$$

By rephrasing the Waldschmidt constant as a solution as a LP, one can prove a Chudnovsky-like result (i.e., a result similar to the statement of Chudnovsky's Conjecture [38]).

Theorem 10.13 ([22, Theorem 5.3]) *Let I be a squarefree monomial ideal and*

$$e = \text{bight}(I) = \max\{\text{ht}(P_i) \mid I = P_1 \cap \cdots \cap P_s\}.$$

Then

$$\widehat{\alpha}(I) \geq \frac{\alpha(I) + e - 1}{e}.$$

Remark 10.14 The above inequality was conjectured to be true for all monomial ideals in Cooper, Embree, Hà, and Hoefel [45].

10.3   Connection to Graph Theory

We end this chapter with a connection between the Waldschmidt constant and graph theory. Recall that we use $G = (V, E)$ to denote a finite simple graph on the vertex set $V = \{x_1, \ldots, x_n\}$ with edge set E. Recall also that the *edge ideal* of

G is defined to be

$$I(G) = \langle x_i x_j \mid \{x_i, x_j\} \in E \rangle.$$

Example 10.15 Let $G = (V, E)$ be the graph with vertex set $V = \{x_1, \ldots, x_5\}$ and edge set $E = \{\{x_1, x_2\}, \{x_2, x_3\}, \{x_3, x_4\}, \{x_4, x_5\}, \{x_5, x_1\}\}$. The graph G is an example of a cycle (specifically, the five cycle) because we can represent it pictorially as in Fig. 10.1. The edge ideal of this graph is then $I(G) = \langle x_1 x_2, x_2 x_3, x_3 x_4, x_4 x_5, x_5 x_1 \rangle$. This ideal is the same ideal as our running example (Example 10.5) in the previous section.

We now introduce a notion that generalizes the idea of a colouring of a graph.

Definition 10.16 Let G be a graph. A *b-fold colouring* of G is an assignment of b colours to each vertex so that adjacent vertices receive different colours. The *b-fold chromatic number* of G, denoted $\chi_b(G)$, is the minimal number of colours needed to give G a b-fold colouring.

Example 10.17 Consider the graph G of Example 10.15. The 2-fold chromatic number of this graph G is $\chi_2(G) = 5$ since the graph can be coloured as in Fig. 10.2. Here, R is RED, B is BLUE, G is GREEN, O is ORANGE, and P is PURPLE.

The b-fold chromatic number allows us to define a new invariant of a graph.

Definition 10.18 The *fractional chromatic number* of G, denoted $\chi_f(G)$, is defined to be

$$\chi_f(G) := \lim_{b \to \infty} \frac{\chi_b(G)}{b}.$$

We can now connect the Waldschmidt constant to the fractional chromatic number.

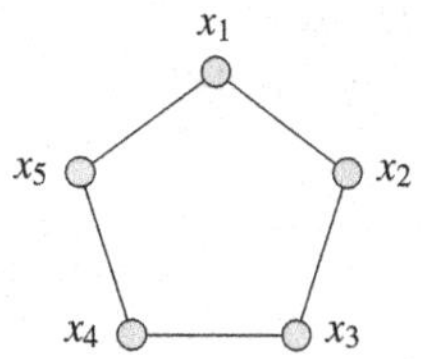

Fig. 10.1 The five cycle graph

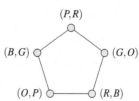

Fig. 10.2 A 2-fold colouring of the five cycle graph

Theorem 10.19 *Let G be a finite simple graph with edge ideal $I(G)$. Then*

$$\widehat{\alpha}(I(G)) = \frac{\chi_f(G)}{\chi_f(G) - 1}.$$

Proof (Sketch of Main Idea) It is known that the fractional chromatic number of a graph can also be expressed as a solution to a LP (see, for example, the book of Scheinerman and Ullman [149]). Then one relates to this LP with the LP of Theorem 10.9.

Remark 10.20 Although we have only stated the above result for edge ideals, the result holds more generally for all squarefree monomial ideals. The appropriate combinatorial object is a hypergraph, that was introduced in Chap. 4. The Waldschmidt constant is then related to the fractional chromatic number of the hypergraph.

Example 10.21 In Example 10.12, we showed that $\widehat{\alpha}(I(C_5)) = \frac{5}{3}$. By Theorem 10.19, we have

$$\frac{5}{3} = \frac{\chi_f(C_5)}{\chi_f(C_5) - 1} \Rightarrow \chi_f(C_5) = \frac{5}{2}.$$

This agrees with [149, Proposition 3.1.2].

Remark 10.22 Besides the Waldschmidt constant, the fractional chromatic number has also appeared in connection to the problems mentioned in Chaps. 1 and 2. In particular, Francisco, Hà, and Van Tuyl [74] used some conditions on the fractional chromatic number to show that a particular prime was an associated prime of a power of a cover ideal.

We have only focused on the case of squarefree monomial ideals. The natural next step is still open:

Question 10.23 Is there a similar procedure to find $\widehat{\alpha}(I)$ for non-squarefree monomial ideals?

Chapter 11
Symbolic Defect

In this chapter we introduce the symbolic defect of a homogeneous ideal. This concept was introduced recently by Galetto, Geramita, Shin, and Van Tuyl [79]. There are a number of interesting questions one can ask about this invariant, and hopefully this chapter will inspire you to investigate the symbolic defect of your favourite family of homogeneous ideals. Throughout this lecture, we will assume that $R = \mathbb{K}[x_1, \dots, x_n]$ is a polynomial ring over an algebraically closed field of characteristic zero, and I will be a homogeneous ideal of R.

11.1 Introducing the Symbolic Defect

We begin with some observations. For any homogeneous ideal I, we always have $I^m \subseteq I^{(m)}$. As a consequence the R-module $I^{(m)}/I^m$ is well-defined. The main idea behind the symbolic defect of an ideal is that $I^{(m)}/I^m$ is somehow a measure of the "failure" of I^m to equal $I^{(m)}$. That is, the "bigger" the module $I^{(m)}/I^m$, the more I^m fails to equal $I^{(m)}$. This suggests we may wish to study the module $I^{(m)}/I^m$ in more detail. Interestingly, there are only a few papers on this module; the papers of which we know include papers by Arsie and Vatne [4], Herzog [104], Herzog and Ulrich [108], Huneke [116], Schenzel [152], and Vasconcelos [165].

But what do we mean by "bigger"? Note that when I is a homogeneous ideal, the R-module $I^{(m)}/I^m$ is also a graded R-module (and also an R/I^m-module). Furthermore, since R is Noetherian, the module $I^{(m)}/I^m$ is Noetherian. Consequently, the quotient $I^{(m)}/I^m$ is a finitely generated graded R-module, and furthermore, the number of minimal generators is an invariant of $I^{(m)}/I^m$. So, one way to measure "bigger" is determine the number of minimal generators of $I^{(m)}/I^m$.

For any R-module M, let $\mu(M)$ denote the number of minimal generators of M. We can then define the symbolic defect of an ideal.

E. Carlini et al., *Ideals of Powers and Powers of Ideals*, Lecture Notes of the Unione Matematica Italiana 27, https://doi.org/10.1007/978-3-030-45247-6_11

Definition 11.1 Let I be a homogeneous ideal of R, and $m \geq 1$ any positive integer. The *m-th symbolic defect* of I, is

$$\text{sdefect}(I, m) := \mu\left(I^{(m)}/I^m\right).$$

The *symbolic defect sequence* of I is the sequence

$$\{\text{sdefect}(I, m)\}_{m \in \mathbb{N}}.$$

Note that it follows directly from the definition that $\text{sdefect}(I, m) = 0$ if and only if $I^{(m)} = I^m$. From this point-of-view, it makes sense to view $\text{sdefect}(I, m)$ as measuring the failure of I^m to equal $I^{(m)}$. Before going further, let's work out an example.

Example 11.2 We consider the monomial ideal

$$I = \langle xy, xz, yz \rangle \subseteq R = \mathbb{K}[x, y, z].$$

Using either a computer algebra system, or computing by hand, we can show

$$I^2 = \langle x^2y^2, x^2y, z, xy^2z, x^2z^2, xyz^2, y^2z^2 \rangle, \quad \text{and}$$
$$I^{(2)} = \langle xyz, x^2y^2, x^2z^2, y^2z^2 \rangle.$$

Thus

$$I^{(2)}/I^2 = \left\langle xyz + I^2, x^2y^2 + I^2, x^2z^2 + I^2, y^2z^2 + I^2 \right\rangle$$
$$= \left\langle xyz + I^2 \right\rangle \subseteq R/I^2.$$

So, $\text{sdefect}(I, 2) = 1$.

We would like to make one other remark about this module since we will return to it at the end of the lecture. Note that the module $I^{(2)}/I^2$ is a graded R-module. We can actually compute the dimension of each graded piece. In particular, we have

$$\dim_{\mathbb{K}}\left[I^{(2)}/I^2\right]_t = \begin{cases} 0 & \text{if } 0 \leq t < 3 \\ 1 & \text{if } t = 3 \\ 0 & \text{if } t > 3. \end{cases}$$

To see why, note that if $t < 3$, then $[I^{(2)}]_t = (0)$, so the first case follows. As we observed above, $I^{(2)}/I^2$ has exactly one generator of degree 3. This gives the result for $t = 3$. For $t > 4$, we claim that $[I^{(2)}]_t = [I^2]_t$. We only need to check that $[I^{(2)}]_t \subseteq [I^2]_t$ since $I^2 \subseteq I^{(2)}$ takes care of the other inclusion. Take any monomial m of degree t in $[I^{(2)}]_t$. Since $I^{(2)}$ is a monomial ideal, m is divisible by one of

$\{xyz, x^2y^2, x^2z^2, y^2z^2\}$. If m is divisible by one of x^2y^2, x^2z^2, or y^2z^2, then it must also be I^2 since these are generators of I^2. Suppose that m is divisible by xyz. Since m has degree $t \geq 4$, there must also be another variable that divides m. But that means that one of x^2yz, xy^2z, or xyz^2 must divide m. So, m is in I^2, as desired.

Now that we have defined the symbolic defect, a number of natural questions arise:

Question 11.3 Let I be a homogeneous ideal of $R = \mathbb{K}[x_1, \ldots, x_n]$.

  (i) How can we compute sdefect(I, m)?
 (ii) When is sdefect$(I, m) = 1$? (In this case, $I^{(m)}$ is "almost" I^m since $I^{(m)} = \langle F \rangle + I^m$ for some homogeneous form F.)
(iii) Are there any applications of sdefect(I, m)?
 (iv) Is the value of sdefect(I, m) related to the containment problem?
  (v) What can one say about the symbolic defect sequence?

In this chapter, we will touch upon $(i) - (iv)$ in Question 11.3. We actually know very little about Question 11.3 (v).

11.2 Some Basic Properties

We quickly describe some basic properties that will be useful for our future discussion.

If sdefect$(I, m) = s$, then there exist s homogeneous forms $F_1, \ldots, F_s \in I^{(m)}$ such that

$$I^{(m)}/I^m = \langle F_1 + I^m, \ldots, F_s + I^m \rangle.$$

Note that this implies that

$$I^{(m)} = \langle F_1, \ldots, F_s \rangle + I^m.$$

It is important to note that the F_i's are not unique. In particular, one can use other coset representatives. That is, for each $i = 1, \ldots, s$, let G_i be a form such that $G_i + I^m = F_i + I^m$. Then

$$I^{(m)}/I^m = \langle G_1 + I^m, \ldots, G_s + I^m \rangle$$

and also $I^{(m)} = \langle G_1, \ldots, G_s \rangle + I^m$. Note, however, that it might be the case that

$$\langle F_1, \ldots, F_s \rangle \neq \langle G_1, \ldots, G_s \rangle.$$

The only thing that is the same is the number of generators. We make this concrete in the next example:

Example 11.4 Let I be as in Example 11.2. Then

$$I^{(2)}/I^2 = \langle xyz + I^2 \rangle = \langle xyz + x^2y^2 + I^2 \rangle$$

but $\langle xyz \rangle \neq \langle xyz + x^2y^2 \rangle$.

The following result summarizes some useful results of sdefect(I, m).

Theorem 11.5 *For all homogeneous radical ideals I,*

(i) sdefect($I, 1$) = 0.
(ii) *if I is a complete intersection,* sdefect(I, m) = 0 *for all $m \geq 1$.*
(iii) *if $\mathbb{X} \subseteq \mathbb{P}^2$, and if $\mathbb{X}$ is not a complete intersection, then* sdefect(I, m) $\neq 0$ *for all $m \geq 2$.*

Proof Statement (i) follows from the fact that $I^{(1)} = I^1$. For (ii), this follows from a classical result of Zariski-Samuel [172, Lemma 5, Apppendix 6] that $I^{(m)} = I^m$ for all $m \geq 1$ when I defines a complete intersection (also see Theorem 9.11). For (iii), see [46, Remark 2.12(i)].

11.3 Computing sdefect(I, m) for Star Configurations

In general, we do not know of any algorithm to compute sdefect(I, m) efficiently.[1] However, one can use the following strategy to compute this value:

Strategy 11.6 (Computing sdefect(I, m)**)** *Let I be a homogeneous ideal of R.*

(a) *Find an ideal J such that $I^{(m)} = J + I^m$.*
(b) *Show that all the minimal generators of J are required.*
(c) sdefect(I, m) = $\mu(J)$.

Note that if one only carries out (a), you have only shown that sdefect(I, m) $\leq$ $\mu(J)$. In [79], Galetto et al. use Strategy 11.6 to find sdefect($I, 2$) when I is a star configuration. Interestingly, the ideal J that we needed for (a) also turns out to be a star configuration. Without further ado, here is the definition of a star configuration.

Definition 11.7 Fix positive integers n, c, and s with $1 \leq c \leq \min\{n, s\}$. Let $\mathscr{L} = \{L_1, \ldots, L_s\}$ be a set of s linear homogeneous polynomials in $\mathbb{K}[x_0, \ldots, x_n]$ such

---

[1]This might be a good research problem.

Fig. 11.1 The linear star configuration of 10 points in $\mathbb{P}^2$

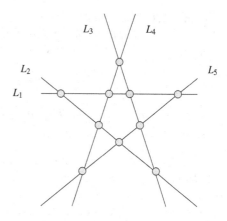

that all subsets of $\mathscr{L}$ of size $c + 1$ are complete intersections. Set

$$I_{c,\mathscr{L}} = \bigcap_{1 \le i_1 < i_2 < \cdots < i_c \le s} \langle L_{i_1}, \ldots, L_{i_c} \rangle.$$

The vanishing locus of $I_{c,\mathscr{L}}$, i.e., $V(I_{c,\mathscr{L}}) \subseteq \mathbb{P}^n$, is a *linear star configuration*.

Remark 11.8 In the above definition, we have required all the elements of $\mathscr{L}$ to be linear forms. One can drop this requirement, and still define a star configuration. To simplify our discussion, we will only focus on the linear case. See [79, Definition 3.1]. For further details on star configurations, see the papers of Geramita, Harbourne, and Migliore [84] and Geramita, Harbourne, Migliore, and Nagel [85].

Example 11.9 The name "star configuration" was suggested by A.V. Geramita in response to Harbourne showing him Fig. 11.1 in 2008 while explaining to him an early version of the results of [44]. Here we have $n = 2$, $s = 5$, and $c = 2$, so we take five linear forms in $\mathbb{K}[x_0, x_1, x_2]$, say $\mathscr{L} = \{L_1, \ldots, L_5\}$. The fact that any three linear forms of $\mathscr{L}$ is a complete intersection is equivalent to the fact that no three of the associated lines meet at the same point. In this case, the star configuration $V(I_{2,\mathscr{L}}) \subseteq \mathbb{P}^2$ is the $10 = \binom{5}{2}$ points of intersections of these five lines. When we draw the five lines, as in Fig. 11.1, we see that they make a "star" shape. Classically, linear star-configurations were sometimes called ℓ-laterals (see, for example, the book of Dogachev [55] just before Lemma 6.3.24).

Example 11.10 Example 11.2 is also an example of a star configuration. In this case, $n = 2$, $s = 3$, and $c = 2$, and the linear forms are $\mathscr{L} = \{x, y, z\}$ in $R = \mathbb{K}[x, y, z]$.

We now describe some properties of the defining ideals of star configurations. In particular, we have an explicit description of the generators of $I_{c,\mathscr{L}}$, and an identity involving the m-th symbolic power of $I_{c,\mathscr{L}}$.

Lemma 11.11 *Fix positive integers n, c, and s with $1 \leq c \leq \min\{n, s\}$ and let $\mathcal{L} = \{L_1, \ldots, L_s\}$ be s linear forms in $R = \mathbb{K}[x_0, \ldots, x_n]$. Then the ideal $I_{c,\mathcal{L}}$ is minimally generated by the forms*

$$\{L_{i_1} \cdots L_{i_{s-c+1}} \mid 1 \leq i_1 < \cdots < i_{s-c+1} \leq s\}.$$

In particular, $I_{c,\mathcal{L}}$ is minimally generated by $\binom{s}{s-c+1}$ homogeneous generators of degree $s - c + 1$.

Proof This result is contained in the work of Park and Shin; in particular see [145, Theorem 2.3] for generation and [145, Corollary 3.5] for minimality.

For the ideal $I_{c,\mathcal{L}}$ of a star configuration, the m-th symbolic power can be computed using the following result. Note that this theorem is reminiscent of the case of squarefree monomials given in Theorem 10.4 (ii). While the next theorem is only stated for linear star configurations, the result holds more generally for all star configurations.

Theorem 11.12 ([85, Theorem 3.6 (i)]) *Let $I_{c,\mathcal{L}}$ be the defining ideal of a linear star configuration in $\mathbb{P}^n$, with $\mathcal{L} = \{L_1, \ldots, L_s\}$. For all $m \geqslant 1$, we have*

$$I_{c,\mathcal{L}}^{(m)} = \bigcap_{1 \leqslant i_1 < \ldots < i_c \leqslant s} \langle L_{i_1}, \ldots, L_{i_c} \rangle^m.$$

The next theorem, due to Geramita, Harbourne, Migliore, and Nagel [85], is an extremely powerful theorem to study the symbolic powers of star configurations. In particular, for some results about symbolic powers of linear star configurations, this theorem implies it is enough to prove the result for the case that each linear form is a variable (and so the defining ideal of the linear star configuration is a monomial ideal). Again, we have specialized a result that holds for star configurations in general.

Theorem 11.13 ([85, Theorem 3.6 (i)]) *Let $I_{c,\mathcal{L}}$ be the defining ideal of a linear star configuration in $\mathbb{P}^n$, with $\mathcal{L} = \{L_1, \ldots, L_s\} \subseteq R = k[x_0, x_1, \ldots, x_n]$. Let $S = k[y_1, \ldots, y_s]$ and define a ring homomorphism $\varphi \colon S \to R$ by setting $\varphi(y_i) = L_i$ for $1 \leq i \leq s$. If I is an ideal of S, then we write $\varphi_*(I)$ to the denote the ideal of R generated by $\varphi(I)$. Let $\mathcal{Y} = \{y_1, \ldots, y_s\}$. Then, for each positive integer m, we have*

$$I_{c,\mathcal{L}}^{(m)} = \varphi_*(I_{c,\mathcal{Y}})^{(m)} = \varphi_*(I_{c,\mathcal{Y}}^{(m)}).$$

With the above machinery, we can now prove the following result.

Theorem 11.14 *Fix positive integers n, c, and s with $1 \le c \le \min\{n, s\}$ and let $\mathscr{L} = \{L_1, \ldots, L_s\}$ be s linear forms in $R = \mathbb{K}[x_0, \ldots, x_n]$. Then for all integers $m \ge 2$,*

$$I_{c,\mathscr{L}}^{(m)} = I_{c,\mathscr{L}}^m + M \text{ for all } m \ge 2$$

where

$$M = \left\langle L_1^{a_1} \cdots L_s^{a_s} \, \middle| \, \begin{array}{l} |\{a_i \mid a_i > 0\}| \ge s - c + 2, \text{ and} \\ a_{i_1} + \cdots + a_{i_c} \ge m \text{ for all } 1 \le i_i < \cdots < i_c \le s \end{array} \right\rangle.$$

Proof To prove this result, it is enough to prove the statement for the case that $\mathscr{L} = \{x_0, \ldots, x_n\}$ in $R = k[x_0, \ldots, x_n]$. We can then apply Theorem 11.13 to prove the general case.

Clearly $I_{c,\mathscr{L}}^{(m)} \supseteq I_{c,\mathscr{L}}^m$. For any monomial $p = x_0^{a_0} \cdots x_n^{a_n} \in M$, we have $a_{i_1} + \cdots + a_{i_c} \ge m$ for all $0 \le i_1 < \cdots < i_c \le n$. Thus $p \in \langle x_{i_1}, \cdots, x_{i_c} \rangle^m$ for all $0 \le i_1 < \cdots < i_c \le n$. Thus $p \in I_{c,\mathscr{L}}^{(m)}$ by Theorem 11.12. Consequently $I_{c,\mathscr{L}}^{(m)} \supseteq M$, and thus $I_{c,\mathscr{L}}^{(m)} \supseteq I_{c,\mathscr{L}}^m + M$.

To show the other containment, consider a monomial $p = x_0^{a_0} x_1^{a_1} \cdots x_n^{a_n} \in I_{c,\mathscr{L}}^{(m)}$. Since $p \in I_{c,\mathscr{L}}^{(m)}$, we have $p \in I_{c,\mathscr{L}}$. Then $|\operatorname{supp}(p)| \ge n - c + 2$ by Lemma 11.11.

If $|\operatorname{supp}(p)| = n - c + 2$, then the complement of $\operatorname{supp}(p)$ in $\{x_0, x_1, \ldots, x_n\}$ has cardinality $c - 1$. Therefore we can write

$$\{x_0, x_1, \ldots, x_n\} \setminus \operatorname{supp}(p) = \{x_{j_1}, \ldots, x_{j_{c-1}}\}.$$

Since $p \in I_{c,\mathscr{L}}^{(m)}$, Theorem 11.12 implies $a_{i_1} + \cdots + a_{i_c} \ge m$ for all $0 \le i_1 < \cdots < i_c \le n$. In particular, for each $x_i \in \operatorname{Supp}(p)$, we must have

$$a_i = a_i + a_{j_1} + \cdots + a_{j_{c-1}} \ge m.$$

Thus p is a multiple of

$$\prod_{x_i \in \operatorname{supp}(p)} x_i^m = \left(\prod_{x_i \in \operatorname{supp}(p)} x_i \right)^m$$

which is the m-th power of a generator of $I_{c,\mathscr{L}}$ by Lemma 11.11. Therefore $p \in I_{c,\mathscr{L}}^m$.

On the other hand, if $|\operatorname{supp}(p)| \ge n - c + 3$, then $p \in M$ by definition.

Example 11.15 Returning to the star configuration of Example 11.2, we have $n = 2, s = 3$, and $c = 2$, with $\mathscr{L} = \{x, y, z\}$. If we consider the case $m = 2$, then the ideal M of Theorem 11.14 is

$$M = \left\langle x^{a_1} y^{a_2} z^{a_3} \;\middle|\; \begin{array}{l} |\{a_i \mid a_i > 0\}| \geq 3 - 2 + 2 = 3, \quad \text{and} \\ a_1 + a_2 \geq 2, a_1 + a_3 \geq 2, a_2 + a_3 \geq 2 \end{array} \right\rangle = \langle xyz \rangle$$

So,

$$I_{2,\mathscr{L}}^{(2)} = \langle xyz \rangle + I_{2,\mathscr{L}}^2.$$

Note that in the above example, the ideal $M = \langle xyz \rangle$ actually equals

$$I_{1,\mathscr{L}} = \langle x \rangle \cap \langle y \rangle \cap \langle z \rangle = \langle xyz \rangle.$$

This is an example of a much more general phenomenon, as first shown in [79, Corollary 3.14].

Corollary 11.16 *With the notation as in Theorem 11.14,*

$$I_{c,\mathscr{L}}^{(2)} = I_{c-1,\mathscr{L}} + I_{c,\mathscr{L}}^2.$$

Proof As in the proof of Theorem 11.14, it is enough to prove the statement for the case that $\mathscr{L} = \{x_0, \ldots, x_n\}$. By [84, Lemma 2.13], we have $I_{c-1,\mathscr{L}} \subseteq I_{c,\mathscr{L}}^{(2)}$, which implies the containment $I_{c,\mathscr{L}}^{(2)} \supseteq I_{c-1,\mathscr{L}} + I_{c,\mathscr{L}}^2$ (these containments hold for any linear star configuration ideal, not just a monomial star configuration ideal). To prove the other containment, we use the fact that our ideals are monomial ideals.

Consider a monomial $p = x_0^{a_0} x_1^{a_1} \ldots x_n^{a_n} \in I_{c,\mathscr{L}}^{(2)}$. As observed in the proof of Theorem 11.14, $|\operatorname{supp}(p)| \geq n - c + 2$ and, in the case of equality, $p \in I_{c,\mathscr{L}}^2$. Assume $|\operatorname{supp}(p)| \geq n - c + 3$. Then p is divisible by one of the generators of $I_{c-1,\mathscr{L}}$ described in Lemma 11.11. Therefore $p \in I_{c-1,\mathscr{L}}$. $\qquad\blacksquare$

We now have enough machinery to determine the symbolic defect for all linear star configurations when $m = 2$.

Theorem 11.17 *Fix positive integers n, c, and s with $1 \leq c \leq \min\{n, s\}$ and let $\mathscr{L} = \{L_1, \ldots, L_s\}$ be s linear forms in $R = \mathbb{K}[x_0, \ldots, x_n]$. Then*

$$\operatorname{sdefect}(I_{c,\mathscr{L}}, 2) = \binom{s}{c-2}.$$

Proof By Corollary 11.16, we know $I_{c,\mathscr{L}}^{(2)} = I_{c-1,\mathscr{L}} + I_{c,\mathscr{L}}^2$, so we need to show that all the generators of $I_{c-1,\mathscr{L}}$ are required. By Lemma 11.11, the ideal $I_{c-1,\mathscr{L}}$ has $\binom{s}{s-(c-1)+1} = \binom{s}{c-2}$ minimal generators of degree $s-c+2$. By the same lemma, the ideal $I_{c,\mathscr{L}}$ is generated in degree $(s - c + 1)$, so $I_{c,\mathscr{L}}^2$ is generated by forms of degree $2(s - c + 1) > s - c + 2$. So, we need all of the generators of $I_{c-1,\mathscr{L}}$ via this degree argument.

The above result leads to the following question:

Question 11.18 What is sdefect($I_{c,\mathscr{L}}, m$) for $m > 2$?

Some upper bounds sdefect($I_{c,\mathscr{L}}, m$) were first found in [79]. Later, Biermann, De Alba, Galetto, Murai, Nagel, O'Keefe, Römer and Seleceanu [17, Remark 4.8] computed sdefect($I_{c,\mathscr{L}}, 3$). Finally, Mantero [134, Corollary 4.12] completely answered the above question; in fact Mantero's work gives an upper bound on the symbolic defect for all star configurations, not just linear star configurations.

We round out this section to by describing two applications of the symbolic defect. We begin by recalling that for any homogeneous ideal $I \subseteq R$, $\alpha(I) = \min\{d \mid (I)_d \neq 0\}$.

Our first application deals with finite sets of points in $\mathbb{P}^2$. That is, let $\mathbb{X} = \{P_1, \ldots, P_s\}$, and let $I_{\mathbb{X}}$ be the associated homogeneous ideal that contains all forms that vanish on $\mathbb{X}$. We say that $\mathbb{X} \subseteq \mathbb{P}^2$ has the *generic Hilbert function* if the Hilbert function of $R/I_{\mathbb{X}}$ is

$$H_{R/I_{\mathbb{X}}}(t) = \min\left\{\dim_{\mathbb{K}} R_t = \binom{t+2}{2}, |\mathbb{X}|\right\} \quad \text{for all integers } t \geq 0.$$

(You should be aware that in some older references, e.g. Geramita, Maroscia, and Roberts [82], a set points is sometimes said to be a set of *points in generic position* if it has the generic Hilbert function. Note that the term "generic" as used in "points in generic position" in this definition is essentially unrelated to the older classical notion of "generic points" mentioned in Remark 8.9.)

As first shown in [79, Theorem 4.6], in some cases, the condition sdefect($I_{\mathbb{X}}, 2$) = 1 forces the points to lie in a special configuration.

Theorem 11.19 *Fix some $\ell \geq 3$, and let $\mathbb{X}$ be a set of $\binom{\ell}{2}$ points in $\mathbb{P}^2$ in generic position. If* sdefect($I_{\mathbb{X}}, 2$) = 1, *then $\mathbb{X}$ is a linear star configuration.*

As a second application, we make the observation that when sdefect(I, m) = 1, then one can create a useful short exact sequence that may be exploited. Specifically, if sdefect(I, m) = 1, this means that there exists a homogeneous form F such that $I^{(m)} = \langle F \rangle + I^m$. We can then build a short exact sequence that relates I^m and $I^{(m)}$:

$$0 \longrightarrow I^m \cap \langle F \rangle \longrightarrow I^m \oplus \langle F \rangle \longrightarrow I^m + \langle F \rangle = I^{(m)} \longrightarrow 0.$$

This short exact sequence allowed [79, Theorem 5.3] to use a mapping cone construction to determine the minimal graded free resolution of $I_{2,\mathscr{L}}^{(2)} \subseteq \mathbb{P}^2$.

11.4 A Connection to the Containment Problem

We now go back to the the containment problem first discussed in Chap. 9 and give a possible link to the symbolic defect. In particular, for any homogeneous ideal $I \subseteq R$, Ein, Lazarsfeld, and Smith [62] showed that for any fixed m, there exists an integer $r \geq m$ such that $I^{(r)} \subseteq I^m$. The containment problem is to find the smallest such r.

Arsie and Vatne [4] observed that the module $I^{(m)}/I^m$ has the following submodules:

$$\frac{I^{(m)}}{I^m} \supseteq \frac{I^{(m+1)} + I^m}{I^m} \supseteq \frac{I^{(m+2)} + I^m}{I^m} \supseteq \cdots \supseteq \frac{I^{(r)} + I^m}{I^m}.$$

The containment problem is thus equivalent to finding the smallest r such that $\frac{I^{(r)}+I^m}{I^m} = 0$. One can ask the following question:

Question 11.20 If sdefect$(I, m) = 1$, does this give any information on the containment problem?

We don't know much about this question. However, here is an example that shows that the containment problem is related to the longest possible chain of proper submodules in $I^{(m)}/I^m$.

Example 11.21 Return to the ideal $I = \langle xy, xz, yz \rangle$ in Example 11.2. We showed that

$$\left[\frac{I^{(2)}}{I^2} \right]_t = 0 \text{ except if } t = 3.$$

In fact, the only possible proper submodule of of $I^{(2)}/I^2$ is the zero module. Since $(I^{(3)} + I^2)/I^2$ is a proper submodule of $I^{(2)}/I^2$, this forces $(I^{(3)} + I^2)/I^2 = 0$, or equivalently, $I^{(3)} \subseteq I^2$.

Chapter 12
Final Comments and Further Reading

The recent survey [50] and the lecture notes of Grifo [88] provide more information on symbolic powers and the containment problem for ideals.

Except in special cases (such as square free monomial ideals), there is no algorithm for computing Waldschmidt constants. Similarly, there is no general algorithm for computing resurgences. See [10] for examples demonstrating some techniques for determining Waldschmidt constants and resurgences, and some specific open problems. For another indication of the difficulty of the problem of computing Waldschmidt constants and resurgences, see [119] for an essentially complete determination of Waldschmidt constants and resurgences for ideals of fat points where the number of points is at most three. This paper also obtains results on symbolic defects in the case of ideals which are not square free. The paper [56] gives additional properties of the symbolic defect, most notably, it is shown that the symbolic defect sequence is a quasi-polynomial function. (We also are pleased to note that the papers [56, 119] are some of the papers resulting from the PRAGMATIC workshop.)

E. Carlini et al., *Ideals of Powers and Powers of Ideals*, Lecture Notes of the Unione Matematica Italiana 27, https://doi.org/10.1007/978-3-030-45247-6_12

Part IV
Unexpected Hypersurfaces

Chapter 13
Unexpected Hypersurfaces

The notion of unexpected hypersurfaces is quite new; research on this topic is growing rapidly but an orderly unified perspective has not yet been achieved. The phenomenon itself can be defined succinctly, but the many examples of unexpectedness that are now known seem to arise in different ways, depending on specific properties available in each context (such as special properties of line arrangements, or of cones, or of characteristic $p > 0$). This currently makes presenting an exposition of reasonable length futile. Thus here we content ourselves with mostly just describing some of the ways unexpectedness arises, with pointers to the literature.

Given a fat point subscheme $X = m_1 p_1 + \cdots + m_r p_r \subset \mathbb{P}^N$, let $V \subseteq R_t = [\mathbb{K}[\mathbb{P}^N]]_t$ be a vector subspace. We say X fails to impose independent conditions on V if

$$\dim_{\mathbb{K}}(V \cap [I(X)]_t) > \max\left\{0, \dim_{\mathbb{K}} V - \sum_i \binom{m_i + N - 1}{N}\right\},$$

where we take $I(p_j)$ to be the ideal generated by all forms (i.e., homogeneous polynomials) vanishing at p_j and set $I(X) = I(p_1)^{m_1} \cap \cdots \cap I(p_r)^{m_r}$, with $[I(X)]_t$ being the vector space span of the forms in $I(X)$ of degree t. It is easy to find examples of a fat point subscheme X of $\mathbb{P}^2$ which fails to impose independent conditions on the space $V = R_t = [\mathbb{K}[\mathbb{P}^2]]_t$ of all forms of degree t. For example, take X to be the fat point scheme $X = 2p_1 + 2p_2$ and take $t = 2$. Then the right hand side of the displayed inequality above is 0 but clearly the square of the linear form defining the line through p_1 and p_2 is a nonzero element of $V \cap [I(X)]_t$, so the left hand side of the displayed equation is positive (and indeed equal to 1).

It is an open problem, even for $N = 2$ with the points p_i being general, to determine all X and t with $V = R_t$ such that X fails to impose independent conditions on V. There is a conjecture in this situation, known as the SHGH

E. Carlini et al., *Ideals of Powers and Powers of Ideals*, Lecture Notes of the Unione Matematica Italiana 27, https://doi.org/10.1007/978-3-030-45247-6_13

Conjecture, which we discuss in more detail below. There has been recent work with V allowed to be certain proper subspaces of R_t. Given some N, t, general points $p_i \in \mathbb{P}^N$ and a subspace $V \subseteq R_t$, it is not very well understood under what circumstances to expect $X = \sum_i m_i p_i \subset \mathbb{P}^N$ to fail to impose independent conditions on V. Thus when it does we will say that the elements of $V \cap [I(X)]_t$ define unexpected hypersurfaces for V, or that V admits an unexpected hypersurface of degree t with respect to X.

13.1 The SHGH Conjecture

Open Problem 13.1 Find all degrees t and integers $m_i > 0$ such that $X = \sum_i m_i p_i \subseteq \mathbb{P}^N$ fails to impose independent conditions on $V = [\mathbb{K}[\mathbb{P}^N]]_t$ when the points p_i are general; i.e.,

$$\dim_{\mathbb{K}}[I(X)]_t > \max\left\{0, \dim_{\mathbb{K}} V - \sum_i \binom{m_i + N - 1}{N}\right\}.$$

The SHGH Conjecture [86, 96, 113, 150], named for the last initials of the authors of the cited papers in temporal order, that is, Segre, Harbourne, Gimigliano, and Hirschowitz, gives a conjectural solution for this when $N = 2$. In the example above, we had $t = 2$ and $X = 2p_1 + 2p_2$, and every element of $[I(X)]_t$ was divisible by L^2 where L was the linear form defining the line through p_1 and p_2.

More generally, let $X = \sum_i m_i p_i \subset \mathbb{P}^2$. We will say that a curve $C \subset \mathbb{P}^2$ is an *exceptional curve* for the points p_i if C is reduced and irreducible with $\deg(C)^2 - \sum_i n_i^2 = -3 \deg(C) + \sum_i n_i = -1$, where $n_i = \mathrm{mult}_{p_i}(C)$. (I.e., n_i is the multiplicity of C at p_i, meaning that that if F is the irreducible form defining C, then $F \in I(p_i)^{n_i}$ but $F \notin I(p_i)^{n_i+1}$.) Note that the line through two distinct points p_1 and p_2 is exceptional for the two points.

Given $X = \sum_i m_i p_i \subset \mathbb{P}^2$, one can show that if the base locus of $[I(X)]_t$ contains a divisor rC with $r > 1$ where C is exceptional for the points p_i (i.e., F^r divides every element of $[I(X)]_t$ where F is the irreducible form defining a curve exceptional for the points p_i), then $\dim_{\mathbb{K}}[I(X)]_t > \max\left\{0, \dim_{\mathbb{K}} R_t - \sum_i \binom{m_i+1}{2}\right\}$.

The SHGH Conjecture says:

Conjecture 13.2 The converse holds when the points p_i are general.

If this is true, then standard techniques allow one to compute $h^0(S, \mathcal{O}_S(F))$ exactly for any divisor F on S where S is the surface obtained by blowing up the points p_i. See section 4 of [98] for more details.

13.2 A More General Problem

The notion of an unexpected hypersurface (introduced by Cook II, Harbourne, Migliore, and Nagel [43] for curves, by Bauer et al. [11] for surfaces, and by Harbourne, Migliore, Nagel, and Teitler [101] for hypersurfaces) was inspired by an example given here as Example 13.7 coming from Di Gennaro, Ilardi, and Vallès [53] (see [53, Figure 2] and the proof of [53, Proposition 7.1]) of 9 points in the plane imposing independent conditions on quartics yet having the property that for each point p there is a quartic curve through the 9 points having multiplicity 3 at p. This is unexpected: the vector space of forms of degree 4 vanishing on Z is 6 dimensional (since the 9 points impose independent conditions on quartics), and one expects that a general point p of multiplicity 3 would impose 6 more conditions, but if this were so there would be no nonzero quartic vanishing on the 9 points and also having a triple point at p. This is a special case of the following problem, which in this generality is wide open:

Open Problem 13.3 Find all integers $t > 0$ and $m_i > 0$ and all fat point subschemes $Z = \sum_j a_j q_j \subset \mathbb{P}^N$ such that $X = \sum_i m_i p_i \subset \mathbb{P}^N$ fails to impose independent conditions on $V = [I(Z)]_t$ where the points $p_i \in \mathbb{P}^N$ are general; i.e.,

$$\dim_{\mathbb{K}}([I(X)]_t \cap V) > \min\left\{0, \dim_{\mathbb{K}} V - \sum_i \binom{m_i + N - 1}{N}\right\}. \qquad (13.1)$$

Example 13.4 Both of the following examples come from Harbourne [98] and show a connection to SHGH. In these examples, X fails to impose independent conditions on $V = [I(Z)]_t$.

(a) If $Z = 0$, this is just is just a case of Problem 13.1, so for $N = 2$ it is conjecturally solved by the SHGH Conjecture.

(b) If $N = 2$ and Z consists of fat points where the points are general, this also in principle is solved by the SHGH Conjecture.

The following example is in characteristic 2; it comes from [43] and is possibly the simplest known example of an unexpected curve.

Example 13.5 Let $N = 2$ with $X = 2p$ for a general point $p \in \mathbb{P}^2$ and with $Z = q_1 + \cdots + q_7$ where the q_i are the 7 points of the Fano plane (so char($\mathbb{K}$) = 2). Then this gives an example of Problem 13.3 with $V = I(Z)_3$, so $t = 3$ and $\dim_{\mathbb{K}} V = 3$. Being singular at p imposes 3 conditions, so we expect no curve, but for every point p there is a cubic form F vanishing on Z and singular at p (specifically $F = a^2 yz(y + z) + b^2 xz(x + z) + c^2 xy(x + y)$ vanishes at the 7 points and is singular at $p = [a : b : c]$).

Example 13.5 is actually the first in a family of similar examples, which heretofore has not been noticed. Let $\mathbb{K}$ be an algebraically closed field of characteristic

$p > 0$. Let $\mathbb{F}$ be a finite subfield, $q = |\mathbb{F}|$. Let Z_q be the reduced scheme of all $q^2 + q + 1$ points of $\mathbb{P}^2_{\mathbb{F}} \subset \mathbb{P}^2_{\mathbb{K}}$.

Theorem 13.6 *Let $p = [a : b : c] \in \mathbb{P}^2_{\mathbb{K}}$ be general, where $\mathbb{K}$ is an algebraically closed field of characteristic $p > 0$. The scheme Z_q admits an unexpected curve (with respect to $X = qp$) for the vector space $V = [I(Z_q)]_{q+1}$ of forms of degree $q + 1$ vanishing on Z_q.*

Proof One first checks that V is the span of $y^q z - y z^q$, $x^q z - x z^q$ and $x^q y - x y^q$. One then checks that $H = (x^q y - x y^q) c^q + (z y^q - y z^q) a^q + (x z^q - z x^q) b^q$ is in V and has a point of multiplicity q at p. But qp imposes $\binom{q+1}{2} \geq 3$ conditions on the 3 dimensional vector space V, hence H defines an unexpected hypersurface.

It is easy to see that $V \subseteq [I(Z_q)]_{q+1}$. To see equality holds, note that Z_q is the union of the complete intersection A of q^2 points defined by $M = \langle x(x^{q-1} - z^{q-1}), y(y^{q-1} - z^{q-1}) \rangle$ with the complete intersection B of $q + 1$ points defined by $N = \langle z, xy(x^{q-1} - y^{q-1}) \rangle$. Thus $I(Z_q) = I(A) \cap I(B) = M \cap N$. A form in N is of the form $Q = Dz + Exy(x^{q-1} - y^{q-1})$ for forms D and E of appropriate degrees. Assume Q is in M. Since $xy(x^{q-1} - y^{q-1}) \in M$ we see that $Dz \in M$. But z does not vanish on A, hence D does, so $D \in M$, hence $Dz \in \langle y^q z - y z^q, x^q z - x z^q \rangle$, so $Q \in \langle y^q z - y z^q, x^q z - x z^q, x^q y - x y^q \rangle$.

Finally, it is clear that $H \in V$, and one checks that H has multiplicity q at p by setting $z = 1$ and translating p to $[0 : 0 : 1]$, and seeing that the result has no terms of degree less than q. $\qquad \blacksquare$

Example 13.7 We now consider the example that motivated the study of unexpectedness. Let $Z \subset \mathbb{P}^2$ be the reduced scheme consisting of the 9 points shown in Fig. 13.1. Namely, start with 4 general points, shown in black, which we may assume are $[0 : 0 : 1]$, $[0 : 1 : 0]$, $[1 : 0 : 0]$, $[1 : 1 : 1]$. There are three singular conics that contain all four general points (one of these conics is shown as a pair of solid lines,

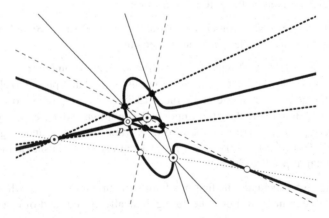

Fig. 13.1 An unexpected quartic curve

another as a bolded pair of dotted lines, and the third as a pair of dashed lines); their singular points are the points shown as dotted circles. Then draw in the line through any two of these three singular points and take the two points (shown as open circles) where this line (shown as a dotted line) intersects the singular conic whose singular point is the third dotted circle. The three singular points and two intersection points give 5 additional points (which specifically are $[0 : 1 : 1]$, $[1 : 0 : 1]$, $[1 : 1 : 0]$, $[-1 : 1 : 0]$, $[1 : 1 : 2]$). Let $X = 3p$ where p is a general point (shown as two concentric circles). Then there is a unique quartic (shown in bold) containing Z with a triple point at p, so $\dim_{\mathbb{K}}([I(X)]_4 \cap V) = \dim_{\mathbb{K}}([I(X + Z)]_4 = 1$, even though here $\dim_{\mathbb{K}}[I(Z)]_4 - 9 = 0$, so the quartic is an unexpected curve for Z with respect to X, and hence gives an example of Problem 13.3.

It is not obvious how to find examples of unexpected curves, but Example 13.7 has a number of special characteristics that have suggested where to look.

For example, as explained in [101], after a projective linear change of coordinates, the 9 points comprising Z consists of the set (up to projective equivalence) of all points $[a : b : c] \in \mathbb{P}^2$ such that all of the coordinates a, b, c are either 0 or ± 1 and either exactly one of them or two of them are nonzero. When these conditions are applied to $(a_0, \ldots, a_N) \in \mathbb{R}^{N+1}$ one obtains the $2(N + 1)^2$ vectors (in $\mathbb{R}^{N+1}$) of the B_{N+1} root system. Regarding these as points in projective space, we get a set $Z_{B_{N+1}}$ of $(N + 1)^2$ points in $\mathbb{P}^N$. Let $p \in \mathbb{P}^N$ be a general point. In addition to the unexpected quartic plane curve given by $Z_{B_{N+1}}$ with $N = 2$ and $X = 3p$, computer testing suggests that $Z_{B_{N+1}} \subset \mathbb{P}^N$ has unexpected cubic hypersurfaces with respect to $X = 3p$ for all $N \geq 5$, and that $Z_{B_{N+1}} \subset \mathbb{P}^N$ has an unexpected quartic hypersurfaces with respect to $X = 4p$ for all $N \geq 3$ (see [101, Table 1]). There is as yet, however, no proof that what the computer testing suggests might always be true is in fact always true. Other root systems also sometimes give unexpected hypersurfaces (see [101, section 3]).

The lines dual to Z for the example given in Example 13.7, which we refer to as the B_3 line arrangement, are also very interesting. After a change of coordinates, they can visualized as shown in Fig. 13.2. This line arrangement is an example of a supersolvable line arrangement.

Let $\mathscr{L}$ be a finite set of 2 or more lines in $\mathbb{P}^2$. Regarding the union $C_{\mathscr{L}}$ of the lines in $\mathscr{L}$ as a plane curve, the singular points of the curve are the points where two or more lines meet. The *multiplicity* of of a singular point is just the number of lines in $\mathscr{L}$ which contain the point. We will denote the number of lines in $\mathscr{L}$ by $d_{\mathscr{L}}$ and by $m_{\mathscr{L}}$ the maximum multiplicity of a singular point of $C_{\mathscr{L}}$.

We say a point Q is a *modular* point of $\mathscr{L}$ if Q is a singular point of $C_{\mathscr{L}}$ with the property that if Q' is any other singular point, then the line $L_{QQ'}$ through Q and Q' is a line in $\mathscr{L}$. (Thus a modular point can see all other singular points by looking along lines of the arrangement.) We say $\mathscr{L}$ is *supersolvable* if it has a modular point. The point at the center of Fig. 13.2 is modular, so the arrangement is supersolvable.

If the lines dual to the points of $Z \subset \mathbb{P}^2$ give a supersolvable line arrangement, it is an interesting question whether Z has unexpected curves with respect to some X. In the case that $X = mp$ and $t = m + 1$, Di Marca, Malara, and Oneto [54]

Fig. 13.2 The nine lines of
the arrangement B_3 (the line
$z = 0$ at infinity is not shown)

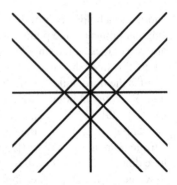

completely characterizes which supersolvable line arrangements are projectively
dual to a point set Z having an unexpected curve. Here is their theorem.

Theorem 13.8 *Let* $Z_{\mathscr{L}} \subset \mathbb{P}^2$ *be the points dual to the lines of a complex
supersolvable line arrangement* $\mathscr{L}$. *Let* p *be a general point. Then* $Z_{\mathscr{L}}$ *admits an
unexpected curve of degree* $d = m + 1$ *with respect to* $X = mp$ *for some* m *if and
only if* $2m_{\mathscr{L}} < d_{\mathscr{L}}$, *in which case* $Z_{\mathscr{L}}$ *has an unexpected curve of degree* $d = m_{\mathscr{L}}$.

13.3   Unexpected Curves and BMSS Duality

Another aspect of Example 13.7 led to the paper [43]. For this example $Z \subset \mathbb{P}^2$ is
a reduced set of points (meaning $Z = \sum_j q_j$, so each point q_j has coefficient 1),
$X = mp$ for a general point $p \in \mathbb{P}^2$ and $t = m + 1$. It is a very interesting question
for which such Z and m, $V = [I(Z)]_t$ has an unexpected curve with respect to
X. The paper [43] gives computationally effective criteria for testing whether $V =
[I(Z)]_t$ admits an unexpected curve with respect to X. When it does, an interesting
phenomenon which [101] calls BMSS duality occurs. This is a reference to the
observation in [11] that the form defining the unexpected curve shown in Fig. 13.1
is actually bihomogeneous.

Let's denote the curve by C_p, so C_p is a quartic curve containing the 9 points
of Z with a triple point at the general point p. We can regard the unexpected curve
as defining a divisor $D \subset \mathbb{P}^2 \times \mathbb{P}^2$, D being the closure of the set of all points
$(p, q) \in \mathbb{P}^2 \times \mathbb{P}^2$ such that C_p is reduced and irreducible and $q \in C_p$. The form
$F_D(a, b, c, x, y, z) \in \mathbb{K}[\mathbb{P}^2 \times \mathbb{P}^2] = \mathbb{K}[a, b, c, x, y, z]$ defining D turns out to be

$$F_D(a, b, c, x, y, z) = c^3 x^3 y - c^3 x y^3 - b^3 x^3 z + (3ab^2 - 3ac^2)x^2 yz +$$

$$(-3a^2 b + 3bc^2)xy^2 z + a^3 y^3 z + (3a^2 c - 3b^2 c)xyz^2 + b^3 xz^3 - a^3 yz^3.$$

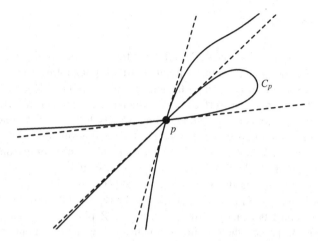

Fig. 13.3 The unexpected quartic curve C_p defined by $F_D(a, b, c, x, y, z) = 0$ (graphed as a solid line) and the graph of $F_D(x, y, z, a, b, c) = 0$ (dashed lines) for $p = [-6 : -5 : 4]$ defining the tangent cone of C_p at p

We see it is homogeneous in both sets of variables, (a, b, c) and (x, y, z), separately (where the coordinates of p are given by (a, b, c) and the coordinates of points q on C_p are given by assigning values to (x, y, z)) and that F_D has degree 4 in (x, y, z) and degree 3 in (a, b, c) (see [101, Example 4.6]). Thus, after picking a specific point $q = [q_0 : q_1 : q_2]$ we get a cubic curve defined by the form $F_D(a, b, c, q_0, q_1, q_2) = 0$, and it turns out that this cubic consists of three lines meeting at q. When $p = q$, these three lines comprise the tangent cone of C_p at p, that is the tangent lines to the three branches of C_p at the quartic curve's triple point p. Thus graphing $F_D(a, b, c, x, y, z) = 0$ for a specific choice $[a : b : c]$ of p gives the quartic curve, and, swapping the variables, graphing $F_D(x, y, z, a, b, c) = 0$ for the same choice of $[a : b : c]$ gives the tangent cone to C_p at p (see Fig. 13.3, which originally appeared as [101, Figure 3]).

We can see this also in the case of Example 13.5. Here Z consists of 7 points, and C_p is a cubic with a double point at p. It gives a divisor $D \subset \mathbb{P}^2 \times \mathbb{P}^2$ defined by the bihomogeneous form $F_D = a^2 yz(y + z) + b^2 xz(x + z) + c^2 xy(x + y)$ of bidegree $(2, 3)$, and one can check that the tangent cone to C_p for each $p = (p_0, p_1, p_2)$ is indeed given by $a^2 p_1 p_2(p_1 + p_2) + b^2 p_0 p_2(p_0 + p_2) + c^2 p_0 p_1(p_0 + p_1)$. Thus the bi-degree, as before, is (m, t) where t is the degree of C_p and $m = \text{mult}_p(C_p)$.

Results of [101] show that this is a fairly general phenomenon, but it is not yet clear to what extent this phenomenon holds. See [60, 101] for further discussion.

13.4   Cones

The unexpected quartic surface (let's call it S_p) for the 16 points Z coming from the root system B_4 and having a general point p of multiplicity 4 is a cone with vertex p, since it consists of the set of lines through p and any point $q \neq p$ where $q \in S_p$.

One way to obtain a hypersurface in $\mathbb{P}^N$ which is a cone having a vertex p which is a general point is to take any codimension 2 variety $V \subset \mathbb{P}^N$. The cone $C_p(V)$ is the set of all lines through p and each point of V. In this situation $V \subset C_p(V)$ for every p, and in fact $V = \cap_{p \in \mathbb{P}^N} C_p(V)$. One can pick a finite set of points $Z \subset V$ so that the $C_p(V)$ is the hypersurface of degree $d = \deg(V)$ having p be a point of multiplicity d, one just needs to choose Z so that $[I(Z)]_d = [I(V)]_d$.

The quartic surface S_p is a cone, but not of the type $C_p(V)$, since one can show that $\cap_{p \in \mathbb{P}^3} S_p$ consists exactly of the 16 points of Z plus 8 additional points (see [101, Section 3.2]); i.e., there is no codimension 2 variety V as in the previous paragraph. However, one can ask if cones of the form $C_p(V)$ ever are unexpected for some finite subset $Z \subset V$. The following result shows that quite often they are.

Theorem 13.9 *Let V be a reduced, equidimensional, non-degenerate subvariety of $\mathbb{P}^N$ ($N \geq 3$) of codimension 2 and degree d (V may be reducible and/or singular but note that $d \geq 2$ since V is non-degenerate, with V being two codimension 2 linear spaces if $d = 2$). Let $p \in \mathbb{P}^N$ be a general point. Choose $Z \subset V$ so that $[I(Z)]_d = [I(V)]_d$. Then $C_p(V)$ is an unexpected hypersurface for Z of degree d and multiplicity d at P. It is the unique unexpected hypersurface of this degree and multiplicity.*

Proof This is essentially just [101, Proposition 2.4] and follows from it immediately using $[I(Z)]_d = [I(V)]_d$.

The bihomogeneous form $F(a, x) \in \mathbb{K}[\mathbb{P}^N \times \mathbb{P}^N] = \mathbb{K}[a_0, \ldots, a_N, x_0, \ldots, x_N]$ defining $C_p(V) \subset \mathbb{P}^N \times \mathbb{P}^N$ has bidegree (d, d). Being a cone, it is its own tangent cone at p. At least when $F(a, x)$ is irreducible it satisfies BMSS duality in the sense that $F(x, a) = \pm F(a, x)$ (see [101, Example 4.1]).

Chapter 14
Final Comments and Further Reading

The papers [43, 101] are essential reading for this section; see also [11] and [60]. The references in these papers give additional papers that may be useful to look at. This research topic is very new but of growing interest, so there are a lot of possible unexplored directions to take.

Two such directions have been taken by two PRAGMATIC work groups. The results of these groups are written up in [54] and [72]. The paper [54] classifies all examples of point sets $Z \subset \mathbb{P}^2$ which have unexpected curves of degree $t = m + 1$ with a general fat singular point $X = mp$, under the assumption that the lines dual to the points of Z comprise what is known as a supersolvable line arrangement. The paper [72] shows that the 9 point set Z in Fig. 13.1 is the only one giving an unexpected quartic with a general point of multiplicity 3.

E. Carlini et al., *Ideals of Powers and Powers of Ideals*, Lecture Notes of the Unione Matematica Italiana 27, https://doi.org/10.1007/978-3-030-45247-6_14

Part V
Waring Problems

Chapter 15
An Introduction to the Waring Problem

An ubiquitous theme in mathematics is the rewriting of mathematical objects. This is usually done to reveal underlying properties, to classify, to solve problems or just for aesthetic reasons!

The first basic example is about a natural number $n \in \mathbb{N}$. There is nothing wrong about n itself, but we all know a fundamental result telling us that n can be uniquely written as a product of primes. For example, $21 = 3 \times 7$. Similarly, we have *Lagrange's Four Squares Theorem* which tells us that any $n \in \mathbb{N}$ can be written as the sum of (at most) four squares of natural numbers. For example, $21 = 4^2 + 2^2 + 1^2$.

Moving from numbers to an array of numbers, we can consider questions about rewriting matrices. A well known case is the so called *rank one decomposition* or *singular values decomposition* in which we want to rewrite a matrix M as a sum of rank one matrices. For example,

$$\begin{bmatrix} 1 & 2 & 3 \\ 4 & 5 & 6 \\ 7 & 8 & 9 \end{bmatrix} = \begin{bmatrix} 1 & 2 & 3 \\ 1 & 2 & 3 \\ 1 & 2 & 3 \end{bmatrix} + \begin{bmatrix} 0 & 0 & 0 \\ 3 & 3 & 3 \\ 6 & 6 & 6 \end{bmatrix}.$$

Note that the rank of the matrix M can be defined as follows:

$$\mathrm{rk}(M) = \min\{r \mid M = M_1 + \cdots + M_r \text{ where } M_i \text{ has rank one}\}.$$

In a very natural way, we can now consider rank one decompositions of *symmetric matrices* and this leads us to consider *quadratic forms*:

$$q(x, \ldots, x_n) = \begin{bmatrix} x_1 & \ldots & x_n \end{bmatrix} S \begin{bmatrix} x_1 \\ \vdots \\ x_n \end{bmatrix}$$

© The Editor(s) (if applicable) and The Author(s), under exclusive licence to Springer Nature Switzerland AG 2020
E. Carlini et al., *Ideals of Powers and Powers of Ideals*, Lecture Notes of the Unione Matematica Italiana 27, https://doi.org/10.1007/978-3-030-45247-6_15

where S is a symmetric $n \times n$ matrix. We can write S in diagonal form $S = PDP^{\top}$ where $D = (d_{ij})$ is a diagonal matrix and P is invertible. If we set

$$
\begin{bmatrix} y_1 \\ \vdots \\ y_n \end{bmatrix} = P^{\top} \begin{bmatrix} x_1 \\ \vdots \\ x_n \end{bmatrix},
$$

then we can rewrite the quadratic form as follows:

$$
q(x_1, \ldots, x_n) = d_{11} y_1^2 + \cdots + d_{nn} y_n^2.
$$

Thus we can rewrite a quadratic form, that is a degree two homogeneous polynomial, q as a sum of squares. Since there is nothing really special about degree two forms, we can also consider forms of degree larger then two and look for *sum of powers decompositions*, (see the survey of Geramita [80] and the paper of Ranestad and Schreyer [147]). For example, if we let

$$
f(x_1, x_2) = x_1^3 + 3x_1^2 x_2 + 3x_1 x_2^2 + x_2^3
$$

we can easily find a (very simple) sum of power decomposition, that is

$$
f(x_1, x_2) = (x_1 + x_2)^3.
$$

We can also look for more general rewriting of homogeneous polynomials, as in Carlini's paper [27]. For example, the polynomial

$$
g(x_1, x_2, x_3, x_4) = (x_1 + x_2)^4 + (x_1 + x_2)(x_3 + x_4)^3 + (x_3 + x_4)^4
$$

can also be written as

$$
g(x_1, x_2, x_3, x_4) = y_1^4 + y_1 y_2^3 + y_2^4.
$$

In what follows we will focus on homogeneous polynomials, that we will call *forms*. In particular, we will address the study of sums of powers decompositions for forms.

15.1   Waring Problems for Homogeneous Polynomials

Inspired by Lagrange's Four Squares Theorem, in 1770 Edward Waring asked the following question: what is the minimal number of d-th powers needed to write any natural number? More explicitly, Waring asked to compute

$$
g(d) = \min\{r \mid \text{ for all } n \in \mathbb{N}, n = n_1^d + \cdots + n_r^d, n_i \in \mathbb{N}\}.
$$

For example, $g(2) = 4$ by Lagrange's Four Squares Theorem, and we know that $g(3) = 9$ and $g(4) = 19$ (and not 16 as we could have hoped).

Similarly, we define

$$G(d) = \min\{r \mid \text{there exists } n_0 \text{ such that if } n \geq n_0, n = n_1^d + \cdots + n_r^d, n_i \in \mathbb{N}\}.$$

The idea is simple: even if $g(2) = 4$, only a finite number of natural numbers *really* need four squares, so that *eventually* three or less squares will suffice. Roughly speaking, $G(d)$ is the number of d-th powers needed to represent large enough integers, while $g(d)$ works for *all* natural numbers.

To compute g is called the *small Waring problem*, while to compute G is called the *big Waring problem*, see [80]. Note that even the existence of these numbers, which was proved by Hilbert, is not trivial.

From now on we are going to focus on the *Waring problem for homogeneous polynomial*, thus we fix some notation:

Definition 15.1 Let $S = \mathbb{K}[x_0, \ldots, x_n]$ be the polynomial ring over the field $\mathbb{K}$, and let S_d be the $\mathbb{K}$ vector space of homogeneous polynomial of degree d.

Definition 15.2 If $F \in S_d$, then a *sum of powers decomposition* of F is an expression of the form

$$F = \lambda_1 L_1^d + \cdots + \lambda_r L_r^d$$

where $\lambda_i \in \mathbb{K}$ and $L_i \in S_1$ for all i.

Note that, if $\mathbb{K} = \mathbb{C}$, then all the coefficients λ_i can be assumed to be equal 1; while, if $\mathbb{K} = \mathbb{R}$, then we can assume $\lambda_i = \pm 1$.

Following the Waring problems for integers we define:

Definition 15.3 For n, d non-negative integers, we let

$$g(n, d) = \min\{r \mid \text{for all } F \in S_d, F = \lambda_1 L_1^d + \cdots + \lambda_r L_r^d \text{ where } \lambda_i \in \mathbb{K}, \ L_i \in S_1\}$$

and

$$G(n, d) = \min\left\{r \;\middle|\; \begin{array}{l} \exists U \subseteq \mathbb{P}(S_d) \text{ open and not empty} \\ \text{such that } \forall F \in U, F = \lambda_1 L_1^d + \cdots + \lambda_r L_r^d \end{array}\right\}.$$

Just a few words on "a generic element of S_d". Technically, such an F belongs to a non-empty Zariski open subset $U \subseteq \mathbb{P}(S_d)$. Since U is dense, we can also think of F, as a "random" element of S_d. In other words, if we pick an element of S_d, it will be generic with probability 1. However, U could be strictly contained in $\mathbb{P}(S_d)$, and thus *non-generic* elements exist. Note that $G(n, d)$ is sometime called the *generic rank* for degree d forms in $n + 1$ variables.

We note that $g(n, 1) = 1, g(n, 2) = n+1$, and $g(1, d) = d+1$, but $g(n, 3)$ is still unknown, while we know that $g(2, 3) = 5, g(2, 4) = 7$, see [26, 52]. In general, we do not know $g(n, d)$. However, we know $G(n, d)$ for all values of n and d thanks to a series of results of Alexander and Hirschowitz [1]:

Theorem 15.4 (Alexander-Hirschowitz) *Let n, d be non-negative integers. Then*

$$G(n, d) = \left\lceil \frac{\binom{n+d}{d}}{n + 1} \right\rceil$$

unless $(n, d) = (n, 2), (2, 4), (3, 4), (4, 3), (4, 4)$.

We note that the exceptions of Alexander-Hirschowitz cases are also called *defective* cases. Furthermore, $G(n, 2) = n + 1$, $G(2, 4) = 5$, $G(3, 4) = 10$, $G(4, 4) = 15$, and $G(4, 3) = 8$.

15.2   Existence Questions

From now on we work over the complex numbers, that is, $\mathbb{K} = \mathbb{C}$.

We defined $g(n, d)$ and $G(n, d)$, and we made some remarks about them. However, we did not address a very basic issue: are our definitions well posed, that is, do $g(n, d)$ and $G(n, d)$ exist? To answer this question we proceed as follows. First we give an upper bound for $g(n, d)$, and this is enough to prove its existence (this will be enough to prove the existence of $G(n, d)$ too).

The key idea behind answering the existence question is to show that there exists a vector space basis of S_d formed by d-th powers of linear forms, and this will imply that $g(n, d) \leq \dim_{\mathbb{K}} S_d = N_d = \binom{n+d}{d}$. We begin by recalling some geometric facts; see also Harris's book [102] for more details:

Definition 15.5 The *d-th Veronese map* is $\nu_d : \mathbb{P}(S_1) \to \mathbb{P}(S_d)$ such that $\nu_d([L]) = [L^d]$. The variety $\nu_d(\mathbb{P}(S_1))$ is called *d-th Veronese embedding of* $\mathbb{P}(S_1)$, or simply, the *d-th Veronese variety*.

We note that we can write down explicitly the Veronese map by choosing suitable monomials bases.

Example 15.6 Consider $n = 2$ and $d = 2$, and choose the monomial basis

$$\{x_0, x_1, x_2\}$$

for S_1, and for S_2 choose the basis

$$\{x_0^2, 2x_0x_1, 2x_0x_2, x_1^2, 2x_1x_2, x_2^2\}.$$

Now we can write $\mathbb{P}(S_1) \simeq \mathbb{P}^2$ and $\mathbb{P}(S_2) \simeq \mathbb{P}^5$, and thus we get

$$\nu_2 : \mathbb{P}^2 \to \mathbb{P}^5$$

$$[a : b : c] \mapsto [a^2 : ab : ac : b^2 : bc : c^2].$$

This map is the well known form of the 2-Veronese embedding of $\mathbb{P}^2$ into $\mathbb{P}^5$.

Following our original idea, we want to find

$$L_1, \ldots, L_{N_d} \in S_1$$

such that the linear span

$$\langle v_d(L_1), \ldots, v_d(L_{N_d}) \rangle$$

is the whole of $\mathbb{P}(S_d)$. We then want to exploit the observation that if the linear span $\langle v_d(L_1), \ldots, v_d(L_{N_d}) \rangle$ is strictly contained in $\mathbb{P}(S_d)$, then there exists a hyperplane H containing all the points $v_d(L_i)$ for $1 \leq i \leq N_d$. We use the following lemma.

Lemma 15.7 *The following are equivalent facts:*

- $v_d([L]) \in H \subset \mathbb{P}(S_d)$;
- $v_d^{-1}(H)$ *is a degree d hypersurface in $\mathbb{P}(S_1)$ containing the point $[L]$.*

Proof We just give the idea for $n = 2$ and $d = 2$. In this case $v_2 : \mathbb{P}^2 \to \mathbb{P}^5$ and we denote with $y_0, \ldots, y_5$ the coordinates on $\mathbb{P}^5$. Thus the hyperplane H is the zero locus of

$$\alpha_0 y_0 + \cdots + \alpha_5 y_5$$

for some $\alpha_i \in \mathbb{C}$, and thus $v_d^{-1}(H)$ is the zero locus of

$$\alpha_0 x_0^2 + \alpha_1 x_0 x_1 + \alpha_2 x_0 x_2 + \alpha_3 x_1^2 + \alpha_4 x_1 x_2 + \alpha_5 x_2^2$$

which is clearly a degree 2 curve in the plane, that is, a conic. Similarly, if $L = ax_0 + bx_1 + cx_2$, then $v_2([L])$ is the point with coordinates

$$[a^2 : ab : ac : b^2 : bc : c^2].$$

Thus, $v_2([L]) \in H$ if and only if

$$\alpha_0 a^2 + \alpha_1 ab + \alpha_2 ac + \alpha_3 b^2 + \alpha_4 bc + \alpha_5 c^2 = 0,$$

and this is equivalent to saying that

$$[L] \in v_d^{-1}(H).$$

With this fact in mind we go back to our idea, that is, we want to find N_d points in $\mathbb{P}^n \simeq \mathbb{P}(S_1)$ not lying on a degree d hypersurface.

Theorem 15.8 *There exist points $P_1, \ldots, P_{N_d} \in \mathbb{P}^n$ not on a degree d hypersurface.*

Proof Again, we just give a sketch of the proof for $n = 2$. If $d = 1$, then $N_1 = 3$ and we need to find three points not a line, which is not difficult to do.

If $d = 2$, then $N_2 = 6$ and we need to find six points not on a conic. To do this, choose three not concurrent lines ℓ_1, ℓ_2 and ℓ_3. Now pick six *distinct* points $P_1 \in \ell_1$, $P_2, P_3 \in \ell_2$ and $P_4, P_5, P_6 \in \ell_3$ not lying at the intersection of the lines. We claim that the set $\mathbb{X} = \{P_1, \ldots, P_6\}$ is not contained in any conic. To see this we use *Bezout's Theorem*. Assume for a contradiction that $\mathscr{C} \supset \mathbb{X}$ is a conic, that is, a degree 2 plane curve. Since $\ell_3 \cap \mathscr{C}$ contains at least three points, and this exceeds the product of the degrees, then $\mathscr{C} \supset \ell_3$, an thus $\mathscr{C} = \ell_3 \cup \ell$ for some line ℓ. Note that ℓ must contain P_1, P_2 and P_3. Thus, $\ell \cap \ell_2$ contains at least two points and, by Bezout's Theorem, $\ell = \ell_2$. Thus, $\mathscr{C} = \ell_3 \cup \ell_2$ and $\mathscr{C}$ does not contain P_1, giving us the contradiction $\mathscr{C} \not\supset \mathbb{X}$.

The same line-by-line construction works for any degree d in $\mathbb{P}^2$ and can be extended to $\mathbb{P}^n$.

We can now prove the existence of $g(n, d)$.

Corollary 15.9 *For non-negative integers n and d, $g(n, d)$ is well defined, and we have that*

$$g(n, d) \le N_d.$$

Proof Choose points $P_i = [L_i] \in \mathbb{P}(S_1)$, $1 \le i \le N_d$ not lying on any degree d hypersurface. Thus, the linear span

$$\langle v_d(L_1), \ldots, v_d(L_{N_d}) \rangle$$

does not lie on any hyperplane, and thus it coincides with the whole of $\mathbb{P}(S_d)$.

Note that for $n = 2$ and $d = 3$, we get $N_3 = 10$, although we know that $g(2, 3) = 5$. Thus the upper bound provided by our argument is quite rough in general. Using a Bertini type argument it is possible to show that $g(n, d) \le N_d - 1$; note that this bound is sharp for $n = 1$. However, the bound is sharp for $n = 1$, where $N_d = d + 1$ and $g(1, d) = d + 1$.

We finally show the existence of $G(n, d)$.

Corollary 15.10 *For non-negative integers n and d, $G(n, d)$ is well defined and we have that*

$$G(n, d) \le N_d.$$

Proof Set $U = \mathbb{P}(S_d)$ and note that U is an open non-empty set. If $F \in U$, then F can be written as the sum of at most $g(n, d)$ powers of linear forms.0 Thus $G(n, d) \le g(n, d)$, and the conclusion follows.

Chapter 16
Algebra of the Waring Problem for Forms

16.1 Apolarity

The most effective tool to deal with the Waring problem for forms is the so-called Apolarity Lemma (see Iarrobino and Kanev [117] and the lecture notes of Carlini, Grieve, and Oeding [32]). To introduce the Apolarity Lemma we need to briefly review some notion from apolarity theory, following Geramita [80].

Definition 16.1 Consider the rings

$$S = \mathbb{C}[x_0, \ldots, x_n] \text{ and } T = \mathbb{C}[y_0, \ldots, y_n]$$

and define the *apolarity action* of T on S by extending the following: let $\partial = y_0^{a_0} \cdots y_n^{a_n} \in T_d$ and let $F \in S$, we set

$$\partial \circ F = \frac{\partial^d}{\partial_{y_0}^{a_0} \cdots \partial_{y_n}^{a_n}} F.$$

Note that this definition makes S into a T-module via differentiation and that the elements of T can be see as partial differential operators on S.

One key object is the following:

Definition 16.2 Let $F \in S_d$. The *annihilator of F* is

$$F^{\perp} = \{\partial \in T \mid \partial \circ F = 0\}.$$

The following lemma collects some of the basic properties of the ideal $F^{\perp}$; for a proof see [80].

E. Carlini et al., *Ideals of Powers and Powers of Ideals*, Lecture Notes of the Unione Matematica Italiana 27, https://doi.org/10.1007/978-3-030-45247-6_16

Lemma 16.3 *If $F \in S_d$, then $F^\perp \subseteq T$ is a homogeneous ideal, and $T/F^\perp$ is an artinian Gorenstein ring with socle degree d, that is, the annihilator of 0 in the $T/F^\perp$ has dimension one and it is generated in degree d.*

Note that $F^\perp$ is often, but informally, called the *perp ideal* of F; we recall some of the basic properties in the following exercise.

Exercise 16.4 Prove that, if $F \in S_d$ is not the zero polynomial, then the Hilbert function $HF(T/F^\perp, t) = \dim_\mathbb{K} T_t - \dim_\mathbb{K} \langle F^\perp \rangle_t$ is symmetric. Furthermore, we have $HF(T/F^\perp, 0) = 1 = HF(T/F^\perp, d)$, and $HF(T/F^\perp, t) = 0$ for $t \geq d + 1$.

Example 16.5 We let $F = xy^2 \in S = \mathbb{C}[x, y]$, and we compute $F^\perp \subset T = \mathbb{C}[X, Y]$. Note that the notational difference in S and T is sometimes suggested by good sense (there are only two variables) and sometimes by personal taste (derivations are now denoted with upper case letters, but they could be denoted by Greek letters).

We proceed to compute $F^\perp$ degree by degree. We start by noting that $\langle F^\perp \rangle_0 = \langle 0 \rangle$ (this is always the case unless $F = 0$).

To compute $\langle F^\perp \rangle_1$ we use linear algebra:

$$aX + bY \in \langle F^\perp \rangle_1 \Leftrightarrow (aX + bY) \circ F = ay^2 + 2bxy = 0 \Leftrightarrow a = b = 0$$

and in conclusion $\langle F^\perp \rangle_1 = \langle 0 \rangle$.

To compute $\langle F^\perp \rangle_2$ we proceed similarly:

$$aX^2 + bXY + cY^2 \in \langle F^\perp \rangle_1 \Leftrightarrow (aX^2 + bXY + cY^2) \circ F = 2by + 2cx = 0 \Leftrightarrow b = c = 0$$

and in conclusion $\langle F^\perp \rangle_1 = \langle X^2 \rangle$.

To compute $\langle F^\perp \rangle_3$:

$$aX^3 + bX^2Y + cXY^2 + dY^3 \in \langle F^\perp \rangle_1 \Leftrightarrow (aX^3 + bX^2Y + cXY^2 + dy^3) \circ c = 0 \Leftrightarrow c = 0$$

and in conclusion $\langle F^\perp \rangle_1 = \langle X^3, X^2Y, Y^3 \rangle$.

Hence, since $\langle F^\perp \rangle_4 = T_4$, we conclude that $F^\perp = \langle X^2, Y^3 \rangle$.

We conclude with two results about the annihilator of monomials (see Carlini, Catalisano, and Geramita [31] for a proof of the first lemma) and of binary forms (the latter is usually attributed to Sylvester, see [41] for a proof of the second lemma):

Lemma 16.6 *If $F = x_0^{a_0} \cdots x_n^{a_n}$ with $n \geq 1$ and $a_0 \leq a_i$ for all i, then*

$$F^\perp = \langle X_0^{a_1+1}, \ldots, X_n^{a_n+1} \rangle.$$

Lemma 16.7 *If $F \in \mathbb{C}[x, y]$ is a degree d form, then*

$$F^\perp = \langle g_1, g_2 \rangle,$$

and $\deg g_1 + \deg g_2 = d + 2$.

16.2 The Apolarity Lemma

We now come to the main result of this chapter.

Lemma 16.8 (Apolarity Lemma [117, Lemma 1.31]) *If $F \in S_d$, then the following are equivalent:*

- $F = \lambda_1 L_1^d + \cdots + \lambda_r L_r^d$, *with* $\lambda_i \in \mathbb{C}$ *and* $L_i \in S_1$ *for all* i;
- $F^\perp \supset I(\mathbb{X})$ *where* $I(\mathbb{X})$ *is the ideal of*

$$\mathbb{X} = \{[L_1], \ldots, [L_r]\} \subset \mathbb{P}^n \simeq \mathbb{P}(S_1),$$

that is, $\mathbb{X}$ is a set of r distinct points.

Example 16.9 Since $(xy^2)^\perp = \langle X^2, Y^3 \rangle$ and

$$I = \langle Y^3 - X^3 \rangle \subset (xy^2)^\perp,$$

is the ideal of three distinct points in $\mathbb{P}^1$, we know that xy^2 can be written as a sum of three cubes.

Example 16.10 Let us see, in some very special cases, what we can say about $F \in S_d$ if we know the set $\mathbb{X}$ such that $F^\perp \supset I(\mathbb{X})$.

If $\mathbb{X} = \{[L]\}$, then $F = \lambda L^d$ for some $\lambda \in \mathbb{C}$. Thus, after changing variables, $F = x^d$.

If $\mathbb{X} = \{[L_1], [L_2]\}$, then $F = \lambda_1 L_1^d + \lambda_2 L_2^d$ for some $\lambda_1, \lambda_2 \in \mathbb{C}$. Thus, after changing variables, $F = x^d + y^d$.

If $\mathbb{X} = \{[L_1], [L_2], [L_3]\}$, then $F = \lambda_1 L_1^d + \lambda_2 L_2^d + \lambda_3 L_3^d$ for some $\lambda_1, \lambda_2, \lambda_3 \in \mathbb{C}$. Now there are two cases to consider: (i) the points of $\mathbb{X}$ are collinear, and (ii) the points of $\mathbb{X}$ are *not* collinear. In case (i), after changing variables, we get that $F = F(x, y)$, that is, F is actually a polynomial in two variables. In case (ii), after changing variables, $F = x^d + y^d + z^d$.

The previous example shows that a form in S could actually involve *less than* n variables. To find the least number of variables needed to write down F, we can explicitly use apolarity as shown by the following lemma; for a proof see [28].

Lemma 16.11 *If $F \in S_d$, then the following are equivalent:*

- *there exist $A_1, \ldots, A_m \in S_1$, $m < n$ such that $F \in \mathbb{C}[A_1, \ldots, A_m]$;*
- $F^\perp \supset I(\mathbb{X})$ *where $\mathbb{X} \subset \mathbb{P}^n$ is a set of points lying on a linear space of dimension m;*
- $\dim_{\mathbb{K}} \langle F^\perp \rangle_1 \geq n - m$.

Example 16.12 Let $n = 2$ and consider $F \in S_d$ such that $F^\perp \supset I(\mathbb{X})$ where

$$\mathbb{X} = \{[x - y], [x - z], [y - z]\} \subset \mathbb{P}(S_1),$$

that is, $\mathbb{X} = \{[1 : -1 : 0], [1 : 0 : -1], [0 : 1 : -1]\} \subset \mathbb{P}^2$. Thus, by the Apolarity Lemma, we have that

$$F = \alpha(x - y)^d + \beta(x - z)^d + \gamma(y - z)^d$$

with $\alpha, \beta, \gamma \in \mathbb{C}$. Hence, by setting $A_1 = x - y$ and $A_2 = x - z$, we see that $F \in \mathbb{C}[A_1, A_2]$.

16.3   Waring Rank

We now introduce the notion of *Waring rank*, or simply *rank*, of a form.

Definition 16.13 If $F \in S_d$, then the *Waring rank* of F is

$$\mathrm{rk}(F) = \min\{r \mid F = L_1^d + \cdots + L_r^d, L_i \in S_1, i = 1, \ldots, r\}.$$

The connection with the Apolarity Lemma is straightforward.

Exercise 16.14 Prove that, if $F \in S_d$, then

$$\mathrm{rk}(F) = \min\{r \mid F^\perp \supset I(\mathbb{X}), \ \mathbb{X} \text{ a set of } r \text{ distinct points}\}.$$

Since the ideal of a finite set of points in $\mathbb{P}^1$ is a principal ideal generated by a squarefree form, we have the following corollary.

Corollary 16.15 *If $F \in S_d$ and $n = 1$, then*

$$\mathrm{rk}(F) = \min\{r \mid \langle F^\perp \rangle_r \text{ contains a non-zero squarefree form}\}.$$

Example 16.16 If $F = xy^2$, then $\mathrm{rk}(F) = 3$ since the least degree of a squarefree form of $\langle X^2, Y^3 \rangle$ is three.

Example 16.17 We let $F = xyz$ and we will compute $\mathrm{rk}(F)$. Since $F = \langle X^2, Y^2, Z^2 \rangle$ does not contain linear forms, we immediately see that $\mathrm{rk}(F) \geq 2$ since any set of one or two points lie on line. Thus $\mathrm{rk}(F) \geq 3$. It is also easy to find an upper bound for the rank; note that

$$\langle Y^2 - X^2, Z^2 - X^2 \rangle \subset F^\perp$$

is the ideal of a (complete intersection) set of four points, and thus $\mathrm{rk}(F) \leq 4$. We have to decide whether $\mathrm{rk}(F) = 3$ or $\mathrm{rk}(F) = 4$. We note that $\mathrm{rk}(F) = 3$ if and only if there exists a set of three distinct points $\mathbb{X} \subset \mathbb{P}^2$ such that $I = I(\mathbb{X}) \subset F^\perp$. Since there is no linear form in $F^\perp$, the three points cannot be collinear, and thus $I = (Q_1, Q_2, Q_3)$ where $Q_1, Q_2, Q_3 \in T_2$ are the minimal generators of the ideal.

Thus, $I \subseteq F^\perp$ and both ideals are generated by three degree two elements, and hence, $I = F$ and this is a contradiction. In conclusion, $\mathrm{rk}(F) = 4$.

The previous example can be extended to provide an upper bound for the rank of any monomial. However, to actually compute the rank, some completely non-trivial idea is necessary and one can obtain the following result.

Theorem 16.18 ([31]) *Consider the monomial $F = x_0^{a_0} \cdots x_n^{a_n}$. If $a_0 \leq a_i$ for all i and $n \geq 1$, then*

$$\mathrm{rk}(F) = \prod_{i=1}^{n} (1 + a_i).$$

There are a few cases in which we can *actually* compute the Waring rank of a given form. For example, we have a nice formula in the case of monomials, and we have a nice algorithm in the case of binary forms, using the so-called *Sylvester's algorithm*. Here is the algorithm: let $F \in \mathbb{C}[x, y]$ and compute $F^\perp$ which we know is a complete intersection of the type $F^\perp = \langle g_1, g_2 \rangle$ and assume that $\deg g_1 \leq \deg g_2$. If g_1 is a squarefree element, then the ideal $\langle g_1 \rangle$ is the ideal of a set of $\deg g_1$ distinct points, and it is easy to see that $\mathrm{rk}(F) = g_1$. If g_1 is not squarefree, then we use Bertini's Theorem which assures us that we can find a squarefree element in $F^\perp$ of $\deg g_2$. It is again easy to see that, in this case, $\mathrm{rk}(F) = \deg g_2$.

16.4   A Sketch of a Proof of the Apolarity Lemma

We finish this chapter with a guided sketch of the proof of the Apolarity Lemma. One direction of the lemma is straightforward.

Lemma 16.19 *Let $\mathbb{X} = \{[L_1], \ldots, [L_r]\}$ with $L_i \in S_1$. If $F = \sum_{i=1}^{r} \lambda_i L_i^d$, with $\lambda_i \in \mathbb{C}$, then $I(\mathbb{X}) \subset F^\perp$.*

Proof Let $\partial \in I(\mathbb{X})$ and note that $\partial \circ F = \sum_{i=1}^{r} \lambda_i \partial \circ L_i^d$. If $\deg \partial > d$, then it is clear that $\partial \circ F = 0$. If $\deg \partial \leq d$, then it is easy to check that

$$\partial \circ L^d \text{ is proportional to } \partial(a_0, \ldots, a_n)$$

where $L = a_0 x_0 + \cdots + a_n x_n$. Since, for $L = L_i$, $\partial(a_0, \ldots, a_n) = 0$, the statement follows.

To complete the proof of the Apolarity Lemma, we need to prove the converse of the previous lemma, that is, *if $I(\mathbb{X}) \subset F^\perp$, with $\mathbb{X} = \{[L_1], \ldots, [L_r]\}$, then $F = \sum_{i=1}^{r} \lambda_i L_i^d$, for $\lambda_i \in \mathbb{C}$.*

We present the proof in the form of a series of exercises which, just by looking at the statement, give a clear idea of the development of the argument. For help filling in the gaps we suggest the reader use Geramita's paper[80] or, for a quick proof, we refer to the introduction of Ranestad and Schreyer [147].

Exercise 16.20

1. Consider the apolarity pairing

$$S_d \times T_d \to \mathbb{C}$$

$$(F, \partial) \mapsto \partial \circ F.$$

For $V \subset S_d$, we define

$$V^{-1} = \{\partial \in T_d \mid \partial \circ F = 0 \text{ for all } F \in V\},$$

while for $W \subset T_d$, we define

$$W^{-1} = \{F \in S_d \mid \partial \circ F = 0, \text{ for all } \partial \in W\}.$$

Prove that the apolarity pairing is a perfect paring, that is, for any $F \in S_d$, $\langle F \rangle^{-1} = T_d$ if and only if $F = 0$ and, for any $T \in T_d$, $\langle T \rangle^{-1} = S_d$ if and only if $T = 0$.
2. Prove that, for $W = \langle F^\perp \rangle_d$, one has $W^{-1} = \{\lambda F : \lambda \in \mathbb{C}\}$.
3. Prove that if W_1, W_2 are subspaces of T_d, then one has $(W_1 \cap W_2)^{-1} = W_1^{-1} + W_2^{-1}$.
4. Prove that if $\mathbb{X} = \{[L_1], \ldots, [L_r]\}$, that is X is a set of r points, then $I(\mathbb{X}) = \bigcap_{i=1}^r \wp_i$, where $\wp_i = I(\{[L_i]\})$.
5. Let $\wp_i = I(\{[L_i]\})$ for $1 \le i \le r$. Prove that, if $F^\perp \supset \bigcap_{i=1}^r \wp_i$, then $\langle F^\perp \rangle_d \supset \bigcap_{i=1}^r (\wp_i)_d$ and thus

$$\langle F^\perp \rangle_d^{-1} \subseteq \langle \wp_1 \rangle_d^{-1} + \cdots + \langle \wp_r \rangle_d^{-1}.$$

6. Let $\wp_i = I(\{[L_i]\})$ and prove that $\langle \wp_1 \rangle_d^{-1} = \langle L_i^d \rangle_d$.

Chapter 17
More on the Waring Problem

In this chapter, we continue to explore problems related to the Waring problem introduced in the last two chapters.

17.1 Maximal Waring Rank

Given n and d, we do not know in general how big the Waring rank can actually be. Of course, for any $S \in S_d$ we know that $\mathrm{rk}(F)$ cannot exceed $\dim_{\mathbb{K}} S_d$, but, usually, the rank of F is (far) less than $\dim_{\mathbb{K}} S_d$. Thus we can ask the question: *what is the largest value of* $\mathrm{rk}(F)$ *when F varies in* S_d? With our notation, we are asking for $g(n, d)$.

In the case of **binary forms**, that is $n = 1$, we have the following complete answer.

Lemma 17.1 *If d is a non-negative integer, then $g(1, d) = d$. If F is a binary degree d form, then $\mathrm{rk}(F) = d$ if only if $F = LM^{d-1}$ for some linear forms L and M.*

Proof We know that $F^{\perp} = \langle g_1, g_2 \rangle$ and $\deg g_1 + \deg g_2 = d + 2$. In particular, unless $F = L^d$, the ideal $F^{\perp}$ is generated in degree at most d and thus, by Sylvester's algorithm, the rank of F is at most d. Clearly, $\mathrm{rk}(LM^{d-1}) = d$. Conversely, if $\mathrm{rk}(F) = d$, then $\deg g_1 = 2, \deg g_2 = d$ and g_1 is a square; by a variable change we may assume $g_1 = X^2$. Hence, $F = Y^d + XY^{d-1}$ for a suitable linear form Y and the conclusion follows.

Note that the binary forms having maximal rank are monomials.

Since the binary case is completely known, we consider the three variable case. Here we know that, for each degree d, there are degree d monomials having

E. Carlini et al., *Ideals of Powers and Powers of Ideals*, Lecture Notes of the Unione Matematica Italiana 27, https://doi.org/10.1007/978-3-030-45247-6_17

rank higher than the generic rank, see Carlini, Catalisano, and Geramita [31]. For example, the monomials xy^dz^d have rank $(d+1)^2$ which is strictly higher than the generic rank $G(2, 2d+1)$ as soon as $d \geq 4$. Unluckily, monomials in three variables do not provide example of forms of maximal rank; for example

- $g(2, 3) = 5$ see [130], but the largest rank monomial is xyz with $\mathrm{rk}(xyz) = 4$;
- $g(2, 4) = 7$ see [52], but the largest rank monomial is xyz^2 with $\mathrm{rk}(xyz^2) = 6$.
- $g(2, 5) = 10$ see [52], but the largest rank monomial is xy^2z^2 with $\mathrm{rk}(xy^2z^2) = 9$.

Very little is known about $g(n, d)$ for $n > 3$, see, for example, Buczinki and Teitler [26]. The best bound known that is valid in general is due to Blekermann and Teitler [20], specifically,

$$g(n, d) \leq 2G(n, d).$$

That is, the maximal Waring rank is at most twice the generic rank. However, this bound is not sharp, e.g., $G(1, 3) = 2$, but $g(1, 3) = 3$. The techniques used to prove these results are various and go from basic topological techniques to a deep algebraic study of annihilator ideals under the apolarity action.

17.2 It is More Complex Over the Reals

In the previous chapters we only considered the complex case, but the real case is, of course, of great interest. Thus, we consider a *real* sum of powers decomposition of $F \in S_d$, that is, an expression of the form

$$F = \lambda_1 L_1^d + \cdots + \lambda_r L_r^d$$

where $\lambda_i \in \mathbb{R}$ and L_i are linear forms with real coefficients for all i. In particular, the *real rank* of F is

$$\mathrm{rk}_{\mathbb{R}}(F) = \min\{r \mid F = \lambda_1 L_1^d + \cdots + \lambda_r L_r^d, \lambda_i \in \mathbb{R}, L_i \in \mathbb{R}[x_0, \ldots, x_n]\}.$$

Clearly the real rank $\mathrm{rk}_{\mathbb{R}}(F)$ is always at least the complex rank of F, but we do not know much beside this.

Up to now binary forms have always provided a safe place to start our investigations. Thus we repeat the process and we start with a success, the real rank of binary monomials, see Boij, Carlini, and Geramita [23].

Theorem 17.2 *Let a, b be non-negative integers. Then*

$$\mathrm{rk}_{\mathbb{R}}(x^a y^b) = a + b.$$

However, even for a binary form F, as soon as we leave the monomial case, we do not know $\mathrm{rk}_{\mathbb{R}}(F)$ in general. To see why this is the case, think of Sylvester's algorithm. We now have to find a minimal degree squarefree element of $F^{\perp}$ having *only real roots*! For more on this see Blekermann [19].

If we stick to monomials we can make some progress, even in more than two variables. Note that the complex rank of $F = x^a y^b$ is $\max\{a+1, b+1\}$, thus the complex and the real rank of the monomial F coincide if and only if either $a = 1$ or $b = 1$. It is a nice surprise to see that this fact is not a binary case accident, but a general property as shown by Carlini, Kummer, Oneto, and Ventura [34].

Theorem 17.3 *The complex and the real rank of $x_0^{a_0} \cdots x_n^{a_n}$ coincide if and only if* $\min\{a_0, \ldots, a_n\} = 1$.

Thus we know the real rank of *some* monomials. The first unknown case is for $F = x^2 y^2 z^2$ where $9 < \mathrm{rk}_{\mathbb{R}}(F)$, but we do not know the exact value of the real rank (though we know that $\mathrm{rk}_{\mathbb{R}}(F) \leq 13$ from Carlini and Ventura [30]). The techniques used to get these results are incredibly different and they range from Descartes's rule of sign to the theory of quadratic forms.

17.3 Waring Loci

In [147] Ranestad and Schreyer introduced the *variety of sum of powers* describing all possible sum of powers decompositions of a given form F involving r summands. This variety is denoted $VSP(F, r)$, and it can be described inside a suitable Hilbert scheme of sets of points in projective space or inside suitable Grassmannians.

Each point of $VSP(F, r)$ gives a sum of powers decomposition of F involving r summands and $VSP(F, r)$ itself describe all such possible decomposition in a beautiful concise geometric way. However, it is possible to break up $VSP(F, r)$, providing a coarser, but easier to control, description of all sum of powers decompositions of F of a fixed length.

The *Waring locus* of $F \in S_d$, as introduced by Carlini, Catalisano, and Oneto [33], describes all linear forms which can appear in a minimal sum of powers decompositions of F

$$\mathscr{W}_F = \left\{ [L] \in \mathbb{P}^n \;\middle|\; F = L^d + \sum_{i=1}^{\mathrm{rk}(F)-1} L_i^d, L_i \in S_d \right\}.$$

Similarly one can define the *forbidden locus* of F as the complement of $\mathscr{W}(F)$, that is, $\mathscr{F}_F = \mathbb{P}^n \setminus \mathscr{W}(F)$; also note that F is assumed to essentially involve $n + 1$ variables as defined by Carlini [28], that is, there is no linear differential operator annihilating F.

In general it is not easy to describe the Waring/forbidden locus of a given form F. But there are families of forms for which we have a complete description, such as, binary forms, monomials, and ternary cubics. We briefly summarize the results contained in [33]. We begin with binary forms.

Theorem 17.4 *Let F be a degree d binary form, and let $g \in F^{\perp}$ be an element of minimal degree. Then*

1. *if $\mathrm{rk}(F) < \lceil \frac{d+1}{2} \rceil$, then $\mathcal{W}_F = V(g)$;*
2. *if $\mathrm{rk}(F) > \lceil \frac{d+1}{2} \rceil$, then $\mathcal{F}_F = V(g)$;*
3. *if $\mathrm{rk}(F) = \lceil \frac{d+1}{2} \rceil$ and d is even, then $\mathcal{F}_F$ is finite and not empty; and*
4. *if $\mathrm{rk}(F) = \lceil \frac{d+1}{2} \rceil$ and d is odd, then $\mathcal{W}_F = V(g)$.*

Cases (1) and (2) deal with a form F *not* having generic rank. If the rank of F is smaller than the generic rank, then the decomposition is unique and the Waring locus is a finite set of points. If the rank of F is larger than the generic rank, then the forbidden locus is finite, that is, all but a finite number of forms can be used to *minimally* decompose F. Think of $d = 3$ and of $[F]$ on a tangent line to the twisted cubic curve in $[L]$, all points of the curve, but not $[L]$, can generate a trisecant plane containing $[F]$. Cases (3) and (4) deal with the generic rank case; in the d odd case the generic form F has unique minimal decomposition, hence $\mathcal{W}_F$ is finite.

We now consider monomials.

Theorem 17.5 *If $M = x_0^{d_0} \cdots x_n^{d_n} \in S$, then*

$$\mathcal{F}_M = V(X_0 \cdots X_m) \subset \mathbb{P}^n,$$

where $m = \max\{i \mid d_i = d_0\}$.

This result says that the forbidden locus for monomials is always closed and non-empty. In particular, almost all linear forms can appear in some minimal sum of powers decomposition of a given monomial. This fact can be used to *simultaneously* decompose monomials as shown by Carlini and Ventura [30]. See also Carlini and Chipalkatti [29] for more on simultaneous Waring decompositions.

In all the previous example either $\mathcal{F}_F$ of $\mathcal{W}_F$ is closed. However, this is not always the case, as it is shown by ternary cubic cusps.

Theorem 17.6 *If $F = z^3 + xy^2$, then*

$$\mathcal{W}_F = \{[0 : 0 : 1]\} \cup \{[a : b : 0] \mid a, b \in \mathbb{C} \text{ and } a \neq 0\}.$$

In words, the Waring locus of $z^3 + xy^2$ is given by a point, corresponding to the linear form z^3, and a line minus a point, corresponding to the binary forms in the variables x, y except the forbidden point for xy^2 which is $[y^3]$.

We note that, in all known cases, the forbidden locus of F is never empty, and actually there is a conjecture: *for any form F the forbidden locus is not empty*. The interest of the conjecture lies in the following heuristic remark: the larger the rank of F is, the more minimal decompositions of F will exist and thus the Waring locus will be quite large. However, forbidden points always seems to exist. For example, the plane cubic of maximal rank $F = x^2(xy + z^2)$ has a forbidden locus consisting of exactly one single point, namely $\mathscr{F}_F = \{[x^3]\}$. The techniques used are a mixture of geometry and algebra: on the one hand one studies the geometry of apolar subsets and on the other hand one uses algebraic techniques to study the rank of a form plus the power of a linear form.

Chapter 18
Final Comments and Further Reading

In these short chapters we just started to explore a very large and intriguing field of mathematics. During the school, our focus was on homogeneous polynomials. However, this is just one of the many landmarks of the subject.

A homogeneous polynomial is just an example of a very special kind of *tensor*, namely a *symmetric tensor*. The setting of Waring problems and Waring rank, both from the algebraic and geometric point of view, can be carried over to tensors; a very good reference for this topic is the book of Landsberg [129].

In these chapters we mainly worked on the field of complex numbers. It is also possible to consider other fields such as the reals and finite fields. However, the situation is very different. Indeed, over finite fields the Waring rank is not always well defined. As an example, consider the form xy over the finite field with two elements. Whenever we compute $(ax + by)^2$ only pure squares survive. Moreover, the same definition of $G(n, d)$ gets affected in some way. Over the reals there are *distinct* open non-empty subsets over which the rank is strictly bounded by *distinct* values which are called *typical ranks* (see, for example Comon and Ottaviani [42]).

Part VI
PRAGMATIC Material

Chapter 19
Proposed Research Problems

In this chapter we collect together the projects that were initially presented to the students of PRAGMATIC. Each project was related to the theme of the workshop, i.e., "Powers of ideals and ideals of powers". Many of these questions are open-ended (and perhaps not well-defined). The intention, however, was to give each group of students some initial suggestions to guide their own research.

19.1 Project 1: The Waldschmidt Constant of Monomial Ideals

(This project is related to the material of Chap. 10.) Let I be a homogeneous ideal of $R = \mathbb{K}[x_1, \ldots, x_n]$. For any homogeneous ideal, let $\alpha(I) = \min\{d \mid I_d \neq 0\}$. That is, $\alpha(I)$ is the smallest degree of nonzero generator of I. The m-th symbolic power of I is defined by

$$I^{(m)} := \bigcap_{P \in \mathrm{ass}(I)} (I^m R_P \cap R)$$

where $\mathrm{ass}(I)$ denotes the associated primes of I and R_P denotes the ring R localized at the prime ideal P.

The *Waldschmidt constant* of I, denoted $\widehat{\alpha}(I)$, is defined to be the limit

$$\widehat{\alpha}(I) = \lim_{m \to \infty} \frac{\alpha(I^{(m)})}{m}.$$

The Waldschmidt constant was introduced in the 1970s by Waldschmidt [167]. More recently, Bocci and Harbourne [21] showed that the Waldschmidt constant can be used to find a lower bound on the resurgence of an ideal. However, in general, computing $\widehat{\alpha}(I)$ is difficult problem. This project is related to the following question.

Question 19.1 Suppose that I is a monomial ideal. What is $\widehat{\alpha}(I)$?

When I is squarefree monomial ideal, then this question was answered in [24, 45], and described in Chap. 10. However, little is known about this question when I is not a squarefree monomial ideal. The goal of this project is to find other monomial ideals where one can compute $\widehat{\alpha}(I)$.

Question 19.1 is very open-ended, so here are some possible suggestions to refine this problem:

- Consider nice classes of non-squarefree monomial ideals. One possible family to consider are the lex-segment monomial ideals (you can find the definition in Herzog-Hibi's book *Monomial Ideals* [106].) You may wish to start with the case that I is generated by all the monomials of degree d in $R = \mathbb{K}[x_1, \ldots, x_n]$.
- The polarization of a monomial ideal I, denoted I^{pol}, is a squarefree monomial ideal constructed from the monomial ideal I (see Definition 4.28). In general, the ideal I^{pol} inherits many properties of I. However, it is currently not known how the Waldschmidt constant of these two ideals compare. Can you determine any relationship between these values? If so, you may be able to use [24] to obtain bounds on the Waldschmidt constant for any monomial ideal.
- With respect to the above problem, you may wish to first consider the case that I is generated only in degree two. So, $I = \langle x_{i_1}^2, x_{i_2}^2, \ldots, x_{i_s}^2 \rangle + I(G)$ where $I(G)$ is the edge ideal of some finite simple graph. You may then be able to use some of the properties of edge ideals (see, e.g., [164]).

Some relevant references are [21, 24, 45, 106, 164]. The paper of Cooper et al. [45] might be of interest in that it uses a polyhedron to study symbolic powers of monomials ideals.

19.2 Project 2: The Symbolic Defect of Monomial Ideals

(This project is related to Chap. 11.) Let I be a homogeneous ideal in the polynomial ring $R = \mathbb{K}[x_1, \ldots, x_n]$. For any positive integer m, the R-module $I^{(m)}/I^m$ is a Noetherian module, so it has finite number of generators. Here, $I^{(m)}$ denotes the m-th symbolic power of I. We define the m-th symbolic defect of I to be

$$\mathrm{sdefect}(I, m) = \text{the number of minimal generators of } I^{(m)}/I^m.$$

This definition first appeared recently in [79]. As a consequence, we still do not know a lot about these numbers, and we believe there are a number of interesting questions one could ask about sdefect(I, m). In particular, this project is to understand the following question:

Question 19.2 If I is a (squarefree) monomial ideal, what is sdefect(I, m)?

Given that this is a broad question, we suggest the possible refinements of this question.

- Let I be either the edge ideal or cover ideal of a finite simply graph. What can be said about sdefect(I, s)? Is there a combinatorial interpretation of this number?
- Continuing from the above point, you may want to first consider the case of cover ideals and the value $s = 2$. Using work of Francisco, Hà, and Van Tuyl [76], you can find a irreducible decomposition of $J(G)^2$. The paper Dupont and Villarreal [61] may be useful to describe the minimal generators of $J(G)^{(2)}$.
- Not much is known about the sequence $\{\text{sdefect}(I, m)\}_{m \in \mathbb{N}}$. Can you find a monomial ideal where the sequence is monotonic?

One other paper that has looked at the module $I^{(m)}/I^m$ is the paper of Arsie and Vatne [4] (a former PRAGMATIC project!). One of their results considers $n + 1$ general sets of points in $\mathbb{P}^n$. In this case, the ideal of the general sets of points is also a monomial ideal. This paper may also be useful.

19.3 Project 3: Regularity of Powers of Ideals

(This project is related to Chaps. 4–6.) For any homogeneous ideal $I \subseteq \mathbb{K}[x_1, \ldots, x_n]$, it is known that for $q \gg 0$, the regularity of I^q is described by a linear polynomial, that is, reg(I^q) $= aq + b$ for some constants a and b. We currently do not have a complete understanding of the meaning of the values a and b.

When $I = I(G)$ is the edge ideal of a graph G, our understanding of reg($I(G)^q$) is better, but still far from complete. This suggests the following broad question:

Question 19.3 Let $I = I(G)$ be edge ideal of a graph. What is reg($I(G)^q$) for $q \gg 0$? A strongly related but much less studied question is that for symbolic powers; what can be said about the regularity of symbolic powers, that is, reg($I(G)^{(q)}$, with $q \gg 0$?

There are several directions to approach this question:

- Find general lower and upper bounds for reg($I(G)^q$) for $q \gg 0$, and characterize graphs which obtain these bounds. For instance, it was shown by Beyarslan, Hà and Trung [12] that if $\nu(G)$ denotes the induced matching number of a graph G then for all $q \geq 1$, we have

$$\text{reg}(I(G)^q) \geq 2q + \nu(G) - 1.$$

It was conjectured that $\operatorname{reg}(I(G)^q) \leq 2q + \operatorname{reg}(I(G)) - 2$ for all $q \geq 1$.

- Characterize graphs that give the simplest linear function, i.e., characterize graphs G such that $\operatorname{reg}(I(G)^q) = 2q$ for all $q \gg 0$. Previous works of Francisco, Hà, Van Tuyl, Nevo and Peeva seem to suggest that the necessary and sufficient condition is $\nu(G) = 1$.
- Find lower and upper bounds for $\operatorname{reg}(I(G)^{(q)})$ similar (or not) to those of $\operatorname{reg}(I(G)^q)$. In particular, characterize graphs for which $\operatorname{reg}(I(G)^{(q)}) = 2q$ for all $q \gg 0$.
- Run examples to see if $\operatorname{reg}(I(G)^{(q)}) = \operatorname{reg}(I(G)^q)$ for all $q \gg 0$. If not, then is $\operatorname{reg}(I(G)^{(q)})$ an asymptotic linear function for $q \gg 0$?
- Investigate $\operatorname{reg}(I(G)^{(q)})$ for special classes of graphs for which the symbolic powers $I(G)^{(q)}$ are well understood, for example perfect graphs.

Some references to this problem include the following papers: [2, 3, 15, 122, 143].

19.4 Project 4: Beyond Perfect Graphs

(This project is related to Chap. 2.) A graph G is perfect if both G and its complement G^c do not contain induced cycles of odd length greater than or equal to five. As was first shown in [76], and discussed Chap. 2, the associated primes of the powers of cover ideals of perfect graphs are fairly well understood. For example, we know that the cover ideals $J(G)$ of perfect graphs satisfy the persistence property.

In the graph theory literature, there are a number of generalizations of perfect graphs, including to hypergraphs (see, for example, [132]). This leads to the question:

Question 19.4 Are there families of hypergraphs that generalize perfect graphs and has some of the similar algebraic properties (e.g., do their edge ideals have the persistence property)?

Some ideas to get you started:

- Chordal graphs are examples of perfect graphs. There have been some attempts at generalizing chordal graphs to hypegraphs; for example, see [67]. We are not aware of anyone studying the associated primes of these hypergraphs. There may be enough structure that you can exploit it.
- As was shown in Chap. 2, if $I^k : I = I^{k-1}$ for all $k \geq 2$, then I has the persistence property. It might also be interesting to check that if $J = J(G)$ is the cover ideal of a perfect graph, then $J(G)$ has this property. This would give a new proof for the persistence property of the cover ideals of perfect graphs. (In [76], the proof is much more explicit in the sense that all the elements of $\operatorname{ass}(J(G)^s)$ are calculated for each s.)

19.5   Project 5: Resurgences for Fat Points

(This project is related to Chaps. 8 and 9.) Let $Z = m_1 p_1 + \cdots + m_s p_s$ be a fat point subscheme of $\mathbb{P}^N$, so $I = I(Z) = \cap I(p_i)^{m_i}$. The resurgence $\rho(I)$ is defined to be

$$\rho(I) = \sup \left\{ \frac{m}{r} \;\middle|\; I^{(m)} \not\subseteq I^r \right\}.$$

The values of $\rho(I)$ are known for some Z, mostly reduced [21, 51, 59, 91]. It is of interest to understand what the values the resurgence can take more comprehensively.

- It should be possible to give a complete answer for $\rho(I)$ for all choices of m_i when the points p_i are collinear points of the plane.
- It should be possible to give a complete answer for $\rho(I)$ for all choices of m_i when the points p_i are the coordinate vertices of $\mathbb{P}^N$, at least when $N = 2$.
- It would also be of interest to determine $\rho(I)$ for each possible Z for small numbers of points in the plane, at least when Z is reduced. (See [83, 90] for classifications of the possible Z, according to their Hilbert functions.)

19.6   Project 6: Unexpected Curves

(This project is related to Chap. 13.) The problems here are based on [43, 53, 70]. The SHGH Conjecture [90, 96, 113, 150] classifies all $(s + 1)$-tuples $(t + 1, m_1, \ldots, m_s)$ where a fat point subscheme $X = m_1 p_1 + \cdots + m_s p_s \subset \mathbb{P}^2$ supported at general points p_i fails to impose independent conditions on $V = R_{t+1}$; i.e., such that

$$\dim_{\mathbb{K}}(I(X)_{t+1} \cap V) > \min \left\{ 0, \dim_{\mathbb{K}} V - \sum_i \binom{m_i + 1}{2} \right\}.$$

(Of course, in this situation we have $I(X)_{t+1} \cap V = I(X)_{t+1}$.) In all known cases, the gcd of $I(X)_{t+1}$ is divisible by F^m for some $m > 1$ where F is irreducible and defines a rational curve of some degree d where $d^2 - \sum_p (\mathrm{mult}_p(F))^2 = -1$, hence $m^2(d^2 - \sum_p (\mathrm{mult}_p(F))^2) < -1$. In particular, there are no known cases where $\dim_{\mathbb{K}}[I(X)]_{t+1} = 1$ and $[I(X)]_{t+1}$ defines a reduced irreducible curve.

Now let $Z = n_1 q_1 + \cdots + n_r q_r \subset \mathbb{P}^2$ where the points q_i are distinct and let $V = I(Z)_{t+1}$. Note that $[I(X)]_{t+1} \cap V = [I(X + Z)]_{t+1}$. Then it can happen that

$$\dim_{\mathbb{K}}([I(X)]_{t+1} \cap V) > \min \left\{ 0, \dim_{\mathbb{K}} V - \sum_i \binom{m_i + 1}{2} \right\},$$

even with $\dim_{\mathbb{K}}[I(X + Z)]_{t+1} = 1$ where $[I(X + Z)]_{t+1}$ defines a reduced irreducible curve C, called an unexpected curve. In all known cases, the unexpected curve is rational and we have $X = m_1 p_1, t = m_1 + 1$ and $n_j = 1$ for all j. In these cases we have $(t + 1)^2 - m_1^2 - \sum_j n_j^2 < -1$.

Over the complex numbers, the least that $t + 1$ can be is 4, but only one example is known in this degree (found in [53], described in detail in Example 13.4(d) and shown in Fig. 13.1).

This raises some questions.

• Find other unexpected quartics, so $X = 3p$ where p is general, $Z = q_1 + \cdots + q_r, r \geq 9$, or show there are no others. One approach here may be to look at possible configurations of 8 points, $Z' = q_1 + \cdots + q_8$ (see [83]) and check whether $I(X + Z')_4$ has any base points not in Z' as p varies. The point here is that no matter how the points of Z' are arranged there is always a quartic $Q(Z', p)$ containing Z' with a general triple point p; the issue is whether there is a choice of Z' such that there is an additional point q such that for every choice of p we have $q \in Q(Z', p)$.

• Look for possible examples where X is not a single fat point. There is always a curve of degree $2m$ with 3 points of multiplicity m, one point of multiplicity $m - 1$ and $2m$ simple points. It may be worth making some of these points general to play the role of X, fix the rest to play the role of Z', and see if there is a point q such that every curve of degree $2m$ through $X + Z'$ also vanishes at q.

Chapter 20
The Art of Research

As is standard at PRAGMATIC, the participants were divided into small groups to work on open research problems, based upon their ranked preferences of the problems. In this iteration of PRAGMATIC, we, as instructors, presented a number of open research problems (see the previous chapter) and some suggested approaches. After the initial assignment of projects, we shifted our focus from lecturing to a focus on mentoring the groups. Not only did we suggest how to make progress on their specific projects, but we also gave more general advice on how to do research and how to present the results.

On the fourth day of the workshop, we met with all nine groups in order to help them get started. As was to be expected, for many of the participants, attacking a new research problem was a new experience. As a result, we set aside some extra time at the end of the first week to give some general advice on how to carry out mathematical research. We have included some of this advice below. We hope that this information will help you develop your own mathematical research program in the years to come.

20.1 Jump In!

When attacking a new research problem, we suggest that you start working on it as quickly as possible. Using the tools you currently know, ask yourself what you can do with the problem. In particular, we highly recommend that you do *not* worry about seeking out all of the background material on your problem before starting. You may feel that you need to learn and understand a large quantity of background material. Unfortunately, this could become an infinite regression, with the end result that you never spend any time working on your problem. Part of the fun of mathematical research is working on your own problem, based on your own ideas, and coming to your own understanding of the problem.

© The Editor(s) (if applicable) and The Author(s), under exclusive licence to Springer Nature Switzerland AG 2020
E. Carlini et al., *Ideals of Powers and Powers of Ideals*, Lecture Notes of the Unione Matematica Italiana 27, https://doi.org/10.1007/978-3-030-45247-6_20

Of course, at some point you will need to look carefully at the literature to see what has been done and to some extent how. But it is easier to integrate background material as your own understanding develops. So, rather than wait to start research until you know everything that has been done or until you've learned everything you think you might need to know, we suggest that you learn as you go. When you get stuck, take a look at what has been done in the area. However, don't stop working on your project to read and completely understand a paper; as soon as you get a new idea, go back to working on your project (possibly reading the paper in parallel, but never in lieu of, working on your project).

20.2   How to Read a Paper When You Feel You Must

Here are a few words of advice about reading papers that might be useful. Never let authors hijack your agenda. Never just read a paper, but rather be an active reader: look at the abstract and think about what the authors claim to be doing. Ask yourself questions: does this seem like the right approach? Are the questions really what you would ask? Are the claimed results really the right answers? Read the introduction and look over the bibliography: what's the motivation? Where did the problem come from? Where do the authors go with it? What was known before? Who are the players? What did they do? Page through the paper and find the main results. How are they stated? Do they make sense? Are the theorems plausible, based on what you know? Try to work out examples (or counterexamples!) of the main results. To what extent do the theorems answer the questions the paper poses? What results are missing (that is, what kinds of results would you have expected that don't seem to be included)? How, in a general sense, do the authors get their results? What new concepts do they introduce? If after this it seems worthwhile to go further, try to sketch out proofs of the results you're mainly interested in: do your approaches seem likely to go through? If not, where do they run into an obstruction? How do the authors get around the problems you run into? Can the authors' proofs be made shorter, or clearer? Do there seem to be gaps? Do there seem to be mistakes? What would need to be true to fill the apparent gaps or to fix the possible mistakes? Annotate the paper with your own thoughts, questions and corrections; go back and modify your annotations as your understanding increases. If the paper merited the attention, keep a record of your annotations so that you can go back later and remind yourself of your thinking. In the end, think about how you would have written the paper.

20.3   Do Experiments and Make Examples

Now that you have "jumped in" and want to start work on your problem, what should you do? Of course, your end goal is to understand something you didn't understand before, and hopefully, as a result, prove an interesting theorem. But before you get to prove your theorem, you need to develop your understanding of what's true and what's not so that you have an idea of what your statement should be. At this stage, you want to take your problem and ask yourself how many examples can you make.

Depending upon the problem, computer software may allow you to make lots of examples (maybe thousands!) but certainly try to make at least some examples somehow. You want to take this data, and think of useful ways of presenting it (e.g., making tables) that allow you to look for patterns. For some problems (e.g., suppose your problem is related to finite simple graphs) you should try to do exhaustive searches for small parameters. Note that the process of making these examples will help you understand the objects you want to study. Making good computer code might also be useful for future research projects (so make sure you document it properly).

20.4   Make Guesses

Now that you have examples and data, start looking for patterns, and start guessing what happens. At any given point, you should have a "working" conjecture, that is, a statement you are actively trying to prove or disprove. If you find a proof, great! If you find a counterexample, use it to change your "working" conjecture. For example, do you need to add more hypotheses? Do you need to change the conclusion? Or do you need to throw out your statement and start all over?

You do not need many examples to make a conjecture. If something seems to work for three examples, and it seems to work for random examples, make this your conjecture. Although it is unlikely it is true, you will learn a lot by finding out why your statement is wrong.

20.5   Write Proofs for Special Cases

When doing research, we suggest that you specialize as much as possible. For example, you may be doing something with the edge ideals of graphs. Instead of trying to prove your statement for all graphs, can you prove it for a special family of graphs (e.g., cycles, complete graphs)? Similarly, if you are studying points in projective space, can you prove your result for points on a line? Once you have a good guess and can prove your "working" conjecture for a special subset of cases, we recommend that you write up a clean and complete proof for this special case.

Don't worry that you can only prove something for a small family. The important thing is that you have proved something. Furthermore, after writing out a proof with all of the details, look at it again and see what you actually proved. Your proof may be more general than the statement you started with, or you may see how to generalize your proof to handle larger classes of cases. Or perhaps you can align your hypotheses to fit the proof; e.g., check what property you really need for the cases your proof handles, and extend your result to all families that have this property.

And of course, another advantage of documenting your attempts and writing up your results is that you will be able to more efficiently get back into it after having to step away, and to recreate your arguments in the future. It's very common to now and then not to have time to do research (e.g., because of teaching responsibilities). You do not want to lose or forget your results just because you didn't take the time to write them down.

20.6 Develop Parallel Questions

When tackling a research problem, we suggest you think of related questions. For example, suppose that you are looking at some properties of points in $\mathbb{P}^2$. While you are studying this problem, you can ask yourself: does this property also work for points in $\mathbb{P}^n$? But note that points in $\mathbb{P}^2$ are also an example of a codimension two subscheme in $\mathbb{P}^2$. So, you can ask if your results also hold for codimension two subschemes. If you get stuck on one approach, you can move to a related, but different, problem.

At any given point, you want to be attacking your problem from various angles. When you get stuck on one approach, you can move to a different approach. Not all variations of the problem will be successful, but you will learn something even from the unsuccessful approaches.

20.7 Writing Papers

Finishing the research does not end the need for creativity. The end result of a successful project is writing your research up for publication, and the writing also takes thought, creativity (and flexibility, especially if you have coauthors)! It's not enough that your theorems be true. You also want to make them interesting. Here we suggest that you consider not getting too attached to your initial formulations. Is there a way to simplify the statements of your main theorem? Can you substantially simplify your proofs by avoiding some special cases? Can you give a simple clean statement with a compelling proof by focusing on the main cases, and move complications to subsidiary results? Think about your write up from the point of view of the reader. Is there a way to present your results that would better grab the interest of a reader? Think about what makes your results interesting and compelling and try to present things to highlight what's of interest and what's compelling. Lead off with simple clean statements with clear proofs if you can; perhaps save the more comprehensive but more complicated theorems and proofs for later in your write up. Once you have their attention and interest, readers will be more willing (and able) to stick with you for the complexities that come later!

If you have coauthors, things can become more delicate. Be sensitive not only to what you think will be important to possible readers, but also to what is important to actual coauthors. Try to see things from your coauthors' perspective. No one likes to see their stuff end up on the cutting room floor. Think about how to highlight what your coauthors have done and still keep the write up clear and concise.

20.8 Collaboration

This comment is not directly about how to do research, although it is important. Working together with others on a research problem is an enjoyable experience. To get the most out of the experience, we wanted to make a couple of suggestions.

First, use the strengths of your collaborator(s). For example, if someone in your group is good at computer programming, have them work on developing examples. Second, although you are working together, make sure you take time to work on your own. It is not necessary for you to spend all your time together. As an example, meet together in the mornings to discuss your plans for the day and share ideas. Then allow group members to work individually, if they want, and get back to together later in the day to discuss your successes and failures. And finally, be a good collaborator. This includes listening to the ideas of others, responding to emails from your collaborators in a timely manner, and helping to write up the results (when it comes to write up your results).

20.9 Presenting Your Work

Talking math with others can be really helpful. What you say about what you're doing, and to whom, depends on where you're at in your project. Talking to collaborators or mentors is good at any time. If you talk to a wider audience you risk getting them interested in your project and jumping in themselves to work on it. This is great after you've done what you wanted to do, and can be very helpful for generating new ideas and alternative directions for you and other to work on. But sometimes you may want to keep working on something without others poaching, and that's OK too.

But eventually you'll want to present your results, perhaps in a seminar or a conference. At that point you have to think about what to say. What you say will depend on how much time you have, who is in the audience, and what you want them to get out of your talk. You might be surprised at how little you can get across in a given amount of time; don't try to do more than your audience can absorb! This can be a particular issue if you use slides rather than giving a board talk. You can squeeze an almost unlimited amount of information when using slides, but your audience has a strictly limited ability to absorb information. And if you have too much material when giving a board talk, the result is usually that you run out of time before getting to what you really wanted to cover. (Especially if you give a board talk, avoid sounding overly rehearsed, but know your lines! Know ahead of time how to say what you plan to say. This will help you use your time most efficiently without getting bogged down in extraneous explanations or straying into unplanned exposition.)

Sometimes the thing that you had to work the hardest on is not the thing that the audience will be most interested to hear about. Make sure you give some background so the audience can appreciate your results. Simplify the statements of your results, and possibly reformulate the necessary definitions, to allow you to get across the main ideas in the least technical way possible. Maybe give an example instead of the actual definition or instead of a theorem statement. Always keep in mind that you are in charge. What you say will determine what kinds of questions the audience has. Don't bring up a topic if you don't want the audience to ask about it (and thereby make you use time on extraneous issues). Try to avoid going into detail on something and ending by saying "but you don't need to know all that, all you need to keep in mind is . . .". Think about how your talk would come across to the listener, then recast it to try to make the impression you're aiming for. When someone who attended your talk meets a colleague who couldn't attend, the second one may ask the first what you talked about. Be sure you said something that the first person can tell the second in a sentence or two. It can even be good to work backward: what main ideas or results do you want them to come away with? What do you need to cover to get to these main ideas and results? What's the least technical and most compelling route to get there? And finally, ending a little early is OK, but never go over your allotted time!

References

1. J. Alexander, A. Hirschowitz, Polynomial interpolation in several variables. J. Algebraic Geom. **4**(2), 201–221 (1995)
2. A. Alilooee, A. Banerjee, Powers of edge ideals of regularity three bipartite graphs. J. Commut. Algebra **9**, 441–454 (2017)
3. A. Alilooee, S. Beyarslan, S. Selvaraja, Regularity of powers of edge ideals of unicyclic graphs. Rocky Mountain J. Math. **49**(3), 699–728 (2019)
4. A. Arsie, J.E. Vatne, A note on symbolic and ordinary powers of homogeneous ideals. Ann. Univ. Ferrara Sez. VII (N.S.) **49**, 19–30 (2003)
5. M.F. Atiyah, I.G. Macdonald, *Introduction to Commutative Algebra* (Addison-Wesley Publishing Co., Reading, MA/London/Don Mills, ON, 1969)
6. A. Bagheri, M. Chardin, H.T. Hà, The eventual shape of Betti tables of powers of ideals. Math. Res. Lett. **20**(6), 1033–1046 (2013)
7. A. Banerjee, The regularity of powers of edge ideals. J. Algebraic Combin. **41**(2), 303–321 (2015)
8. A. Banerjee, S. Beyarslan, H.T. Hà, Regularity of powers of edge ideals: from local properties to global bounds (2018). Preprint. arXiv:1805.01434
9. Th. Bauer, S. Di Rocco, B. Harbourne, M. Kapustka, A. Knutsen, W. Syzdek, T. Szemberg, A primer on Seshadri constants, in *Interactions of Classical and Numerical Algebraic Geometry.* Contemporary Mathematics, vol. 496 (American Mathematical Society, Providence, RI, 2009), pp. 33–70
10. Th. Bauer, S. Di Rocco, B. Harbourne, J. Huizenga, A. Seceleanu, T. Szemberg, Negative curves on symmetric blowups of the projective plane, resurgences and Waldschmidt constants. Int. Math. Res. Not. **24**, 7459–7514 (2019)
11. Th. Bauer, G. Malara, T. Szemberg, J. Szpond, Quartic unexpected curves and surfaces. Manuscript Math. (to appear). arXiv:1804.03610
12. S. Bayati, J. Herzog, G. Rinaldo, On the stable set of associated prime ideals of a monomial ideal. Arch. Math. (Basel) **98**(3), 213–217 (2012)
13. E. Bela, G. Favacchio, N. Tran, In the shadows of a hypergraph: looking for associated primes of powers of squarefree monomial ideals. J. Algebraic Combin. (to appear). arXiv:1809.08796
14. C. Berge, *Hypergraphs: Combinatorics of Finite Sets* (North-Holland, New York, 1989)
15. S. Beyarslan, H.T. Hà, T.N. Trung, Regularity of powers of forests and cycles. J. Algebraic Combin. **42**(4), 1077–1095 (2015)
16. A. Bhat, J. Biermann, A. Van Tuyl, Generalized cover ideals and the persistence property. J. Pure Appl. Algebra **218**(9), 1683–1695 (2014)

E. Carlini et al., *Ideals of Powers and Powers of Ideals*, Lecture Notes of the Unione Matematica Italiana 27, https://doi.org/10.1007/978-3-030-45247-6

17. J. Biermann, H. De Alba, F. Galetto, S. Murai, U. Nagel, A. O'Keefe, T. Römer, A. Seceleanu, Betti numbers of symmetric shifted ideals (2019). Preprint. arXiv:1907.04288

18. S. Bisui, H.T. Hà, A.C. Thomas, Fiber invariants of projective morphisms and regularity of powers of ideals Acta Math. Vietnam. (to appear). arXiv:1901.04425

19. G. Blekherman, Typical real ranks of binary forms. Found. Comput. Math. **15**, 793–798 (2015)

20. G. Blekherman, Z. Teitler, On maximum, typical and generic ranks. Math. Ann. **362**, 1021–1031 (2015)

21. C. Bocci, B. Harbourne, Comparing powers and symbolic powers of ideals. J. Algebraic Geom. **19**(3), 399–417 (2010)

22. C. Bocci, S. Cooper, E. Guardo, B. Harbourne, M. Janssen, U. Nagel, A. Seceleanu, A. Van Tuyl, T. Vu, The Waldschmidt constant for squarefree monomial ideals. J. Algebraic Combin. **44**(4), 875–904 (2016)

23. M. Boij, E. Carlini, A.V. Geramita, Monomials as sums of powers: the real binary case. Proc. Am. Math. Soc. **139**, 3039–3043 (2011)

24. M. Brodmann, Asymptotic stability of $\mathrm{Ass}(M/I^n M)$. Proc. Am. Math. Soc. **74**, 16–18 (1979)

25. W. Bruns, J. Herzog, *Cohen-Macaulay Rings*. Cambridge Studies in Advanced Mathematics, vol. 39 (Cambridge University Press, Cambridge, 1993)

26. J. Buczyński, Z. Teitler, Some examples of forms of high rank. Collect. Math. **67**, 431–441 (2016)

27. E. Carlini, Codimension one decompositions and Chow varieties, in *Projective Varieties with Unexpected Properties* (Walter de Gruyter, Berlin, 2005), pp. 67–79

28. E. Carlini, Reducing the number of variables of a polynomial, in *Algebraic Geometry and Geometric Modeling*. Math Vis (Springer, Berlin, 2006), pp. 237–247

29. E. Carlini, J. Chipalkatti, On Waring's problem for several algebraic forms. Comment. Math. Helv. **78**, 494–517 (2003)

30. E. Carlini, E. Ventura, A note on the simultaneous Waring rank of monomials. Illinois J. Math. **61**(3–4), 517–530 (2017)

31. E. Carlini, M.V. Catalisano, A.V. Geramita, The solution to the Waring problem for monomials and the sum of coprime monomials. J. Algebra **370**, 5–14 (2012)

32. E. Carlini, N. Grieve, L. Oeding, Four lectures on secant varieties, in *Connections Between Algebra, Combinatorics, and Geometry*. Springer Proceedings in Mathematics & Statistics, vol. 76 (Springer, New York, 2014), pp. 101–146

33. E. Carlini, M.V. Catalisano, A. Oneto, Waring loci and the Strassen conjecture. Adv. Math. **314**, 630–662 (2017)

34. E. Carlini, M. Kummer, A. Oneto, E. Ventura, On the real rank of monomials. Math. Z. **286**, 571–577 (2017)

35. M. Chardin, Some results and questions on Castelnuovo-Mumford regularity, in *Syzygies and Hilbert Functions*. Lecture Notes in Pure and Applied Mathematics, vol. 254 (Chapman & Hall/CRC, Boca Raton, FL, 2007), pp. 1–40

36. M. Chardin, Powers of ideals and the cohomology of stalks and fibers of morphisms. Algebra Number Theory **7**(1), 1–18 (2013)

37. J. Chen, S. Morey, A. Sung, The stable set of associated primes of the ideal of a graph. Rocky Mountain J. Math. **32**(1), 71–89 (2002)

38. G.V. Chudnovsky, Singular points on complex hypersurfaces and multidimensional Schwarz lemma, in *Seminar on Number Theory*, Paris 1979–1980. Progress in Mathematics, vol. 12 (Birkhäuser, Boston, MA, 1981), pp. 29–69

39. Y. Cid-Ruiz, Regularity and Gröbner bases of the Rees algebra of edge ideals of bipartite graphs. Matematiche (Catania) **73**(2), 279–296 (2018)

40. Y. Cid-Ruiz, S. Jarafi, N. Nemati, B. Picone, Regularity of bicyclic graphs and their powers. J. Algebra Appl. (to appear). arXiv:1802.07202

41. G. Comas, M. Seiguer, On the rank of a binary form. Found. Comput. Mat. **11**(1), 65–78 (2011)

42. P. Comon, G. Ottaviani, On the typical rank of real binary forms. Linear Multilinear Algebra **60**(6), 657–667 (2012)

43. D. Cook II, B. Harbourne, J. Migliore, U. Nagel, Line arrangements and configurations of points with an unusual geometric property. Compos. Math. **154**(10), 2150–2194 (2018)

44. S. Cooper, B. Harbourne, Z. Teitler, Combinatorial bounds on Hilbert functions of fat points in projective space. J. Pure Appl. Algebra **215**(9), 2165–2179 (2011)

45. S. Cooper, R.J.D. Embree, H.T. Hà, A.H. Hoefel, Symbolic powers of monomial ideals. Proc. Edinb. Math. Soc. (2) **60**(1), 39–55 (2016)

46. S. Cooper, G. Fatabbi, E. Guardo, A. Lorenzini, J. Migliore, U. Nagel, A. Seceleanu, J. Szpond, A. Van Tuyl, Symbolic powers of codimension two Cohen-Macaulay ideals (2016). Preprint. arXiv:1606.00935

47. D. Cox, J. Little, D. O'Shea, *Ideals, Varieties, and Algorithms. An Introduction to Computational Algebraic Geometry and Commutative Algebra*, 4th edn. Undergraduate Texts in Mathematics (Springer, Cham, 2015)

48. S.D. Cutkosky, J. Herzog, N.V. Trung, Asymptotic behaviour of the Castelnuovo-Mumford regularity. Composito Math. **118** (1999), 243–261.

49. H. Dao, C. Huneke, J. Schweig, Bounds on the regularity and projective dimension of ideals associated to graphs. J. Algebraic Combin. **38**(1), 37–55 (2013)

50. H. Dao, A. De Stefani, E. Grifo, C. Huneke, L. Núñez-Betancourt, Symbolic powers of ideals, in *Singularities and Foliations. Geometry, Topology and Applications*. Springer Proceedings in Mathematics & Statistics, vol. 222 (Springer, Cham, 2018), pp. 387–432

51. A. Denkert, M. Janssen, Containment problem for points on a reducible conic in P^2. J. Algebra **394**, 120–138 (2013)

52. A. De Paris, The asymptotic leading term for maximum rank of ternary forms of a given degree. Linear Algebra Appl. **500**, 15–29 (2016)

53. R. Di Gennaro, G. Ilardi, J. Vallès, Singular hypersurfaces characterizing the Lefschetz properties. J. Lond. Math. Soc. (2) **89**(1), 194–212 (2014)

54. M. Di Marca, G. Malara, A. Oneto, Unexpected curves arising from special line arrangements. J. Algebraic Combin. (to appear) arXiv:1804.02730

55. I.V. Dolgachev, *Classical Algebraic Geometry: A Modern View* (Cambridge University Press, Cambridge, 2012)

56. B. Drabkin, L. Guerrieri, Asymptotic invariants of ideals with Noetherian symbolic Rees algebra and applications to cover ideals. J. Pure Appl. Algebra (to appear). arXiv:1802.01884

57. P. Dubreil, Sur quelques propriétés des systèmes de points dans le plan et des courbes gauches algébriques. Bull. Soc. Math. France **61**, 258–283 (1933)

58. M. Dumnicki, T. Szemberg, H. Tutaj-Gasińska, Counterexamples to the $I^{(3)} \subset I^2$ containment. J. Algebra **393**, 24–29 (2013)

59. M. Dumnicki, B. Harbourne, U. Nagel, A. Seceleanu, T. Szemberg, H. Tutaj-Gasińska, Resurgences for ideals of special point configurations in $\mathbb{P}^N$ coming from hyperplane arrangements. J. Algebra **443**, 383–394 (2015)

60. M. Dumnicki, Ł. Farnik, B. Harbourne, G. Malara, J. Szpond, H. Tutaj-Gasińska, A matrixwise approach to unexpected hypersurfaces (2019). Preprint. arXiv:1901.03725

61. L. Dupont, R. Villarreal, Vertex covers and irreducible representations of Rees cones. Algebra Discrete Math. **10**(2), 64–86 (2011)

62. L. Ein, R. Lazarsfeld, K. Smith, Uniform behavior of symbolic powers of ideals. Invent. Math. **144**, 241–252 (2001)

63. D. Eisenbud, *Commutative Algebra: with a View Toward Algebraic Geometry*. Graduate Texts in Mathematics, vol. 150 (Springer, New York, 1995)

64. D. Eisenbud, S. Goto, Linear free resolutions and minimal multiplicity. J. Algebra **88**(1), 89–133 (1984)

65. D. Eisenbud, J. Harris, Powers of ideals and fibers of morphisms. Math. Res. Lett. **17**(2), 267–273 (2010)

66. D. Eisenbud, B. Ulrich, Notes on regularity stabilization. Proc. Am. Math. Soc. **140**(4), 1221–1232 (2012)

67. E. Emtander, A class of hypergraphs that generalizes chordal graphs. Math. Scand. **106**(1), 50–66 (2010)

68. N. Erey, Powers of edge ideals with linear resolutions. Commun. Algebra **46**, 4007–4020 (2018)

69. H. Esnault, E. Viehweg. Sur une minoration du degré d'hypersurfaces s'annulant en certains points. Math. Ann. **263**(1), 75–86 (1983)

70. D. Faenzi, J. Vallès, Logarithmic bundles and line arrangements, an approach via the standard construction. J. Lond. Math. Soc. **90**(2), 675–694 (2014)

71. S. Faridi, The facet ideal of a simplicial complex. Manuscripta Math. **109**(2), 159–174 (2002)

72. Ł. Farnik, F. Galuppi, L. Sodomaco, W. Trok, On the unique unexpected quartic in $\mathbb{P}^2$. J. Algebraic Combin. (to appear). arXiv:1804.03590

73. H. Fekete, Über die Verteilung der Wurzeln bei gewissen algebraischen Gleichungen mit ganzzahligen Koeffizienten. Math. Z. **17**, 228–249 (1923)

74. C. Francisco, H.T. Hà, A. Van Tuyl, A conjecture on critical graphs and connections to the persistence of associated primes. Discrete Math. **310**, 2176–2182 (2010)

75. C. Francisco, H.T. Hà, A. Van Tuyl, Associated primes of monomial ideals and odd holes in graphs. J. Algebraic Combin. **32**(2), 287–301 (2010)

76. C. Francisco, H.T. Hà, A. Van Tuyl, Colorings of hypergraphs, perfect graphs, and associated primes of powers of monomial ideals. J. Algebra **331**, 224–242 (2011)

77. C. Francisco, H.T. Hà, J. Mermin, Powers of square-free monomial ideals and combinatorics. *Commutative Algebra* (Springer, New York, 2013), pp. 373–392

78. R. Fröberg, On Stanley-Reisner rings, in *Topics in Algebra, Part 2 (Warsaw, 1988)*. vol. 26 (Banach Center Publication, PWN, Warsaw, 1990), pp. 55–70

79. F. Galetto, A.V. Geramita, Y.-S. Shin, A. Van Tuyl, The symbolic defect of an ideal. J. Pure Appl. Algebra **223**(6), 2709–2731 (2019)

80. A.V. Geramita, Inverse systems of fat points: Waring's problem, secant varieties of veronese varieties and parameter spaces for Gorenstein ideals, in *The Curves Seminar at Queen's*, vol. 10 (1996), pp. 2–114

81. A.V. Geramita, P. Maroscia, The ideal of forms vanishing at a finite set of points in $\mathbb{P}^n$. J. Algebra **90**, 528–555 (1984)

82. A.V. Geramita, P. Maroscia, L. Roberts, The Hilbert function of a reduced k-algebra. J. Lond. Math. Soc. **28**, 443–452 (1983)

83. A.V. Geramita, B. Harbourne, J. Migliore, Classifying Hilbert functions of fat point subschemes in $\mathbb{P}^2$. Collect. Math. **60**, 159–192 (2009)

84. A.V. Geramita, B. Harbourne, J. Migliore, Star configurations in $\mathbb{P}^n$. J. Algebra **376**, 279–299 (2013)

85. A.V. Geramita, B. Harbourne, J. Migliore, U. Nagel, Matroid configurations and symbolic powers of their ideals. Trans. Am. Math. Soc. **369**(10), 7049–7066 (2017)

86. A. Gimigliano, On linear systems of plane curves. Ph.D. Thesis, Queen's University, Kingston (1987)

87. D.R. Grayson, M.E. Stillman, Macaulay2, a software system for research in algebraic geometry. Available at http://www.math.uiuc.edu/Macaulay2/

88. E. Grifo, *Symbolic Powers*. Lecture Notes for RTG Advanced Summer Mini-course in Commutative Algebra, May 2018. Available at: www.math.utah.edu/agtrtg/commutative-algebra/Grifo_symbolic_powers.pdf

89. Y. Gu, H.T. Hà, J. O'Rourke, J. Skelton, Symbolic powers of edge ideals of graphs (2018). Preprint. arXiv:1805.03428

90. E. Guardo, B. Harbourne, Configuration types and cubic surfaces. J. Algebra **320**, 3519–3533 (2008)

91. E. Guardo, B. Harbourne, A. Van Tuyl, Asymptotic resurgences for ideals of positive dimensional subschemes of projective space. Adv. Math. **246**, 114–127 (2013)

92. H.T. Hà, Asymptotic linearity of regularity and a*-invariant of powers of ideals. Math. Res. Lett. **18**(1), 1–9 (2011)

93. H.T. Hà, M. Sun, Squarefree monomial ideals that fail the persistence property and nonincreasing depth. Acta Math. Vietnam. **40**(1), 125–137 (2015)

94. H.T. Hà, A. Van Tuyl, Monomial ideals, edge ideals of hypergraphs, and their minimal graded free resolutions. J. Algebraic Combin. **27**(2), 215–245 (2008)

95. H.T. Hà, D.H. Nguyen, N.V. Trung, T.N. Trung, Symbolic powers of sums of ideals. Math. Z. (2019, to appear). https://doi.org/10.1007/s00209-019-02323-8

96. B. Harbourne, The geometry of rational surfaces and Hilbert functions of points in the plane, in *Proceedings of the 1984 Vancouver Conference in Algebraic Geometry, CMS Conference Proceedings*, vol. 6 (American Mathematical Society, Providence, RI, 1986), 95–111

97. B. Harbourne, Global aspects of the geometry of surfaces. Ann. Univ. Paed. Cracov. Stud. Math. **9**, 5–41 (2010)

98. B. Harbourne, Asymptotics of linear systems, with connections to line arrangements. *Phenomenological Approach to Algebraic Geometry*, vol. 116 (Banach Center Publication, Polish Academy of Sciences Institute of Mathematics, Warsaw, 2018), pp. 87–135

99. B. Harbourne, C. Huneke, Are symbolic powers highly evolved? J. Ramanujan Math. Soc. **28**(Special Issue-2013, 3), 311–330

100. B. Harbourne, J. Roé, Extendible estimates of multipoint Seshadri constants (2013). Preprint. arXiv:math/0309064

101. B. Harbourne, J. Migliore, U. Nagel, Z. Teitler, Unexpected hypersurfaces and where to find them. Mich. Math. J. (to appear). arXiv:1805.10626

102. J. Harris, *Algebraic Geometry: A First Course*. Graduate Texts in Mathematics, vol. 133 (Springer, New York, 1992)

103. R. Hartshorne, *Algebraic Geometry*. Graduate Texts in Mathematics, vol. 52 (Springer, New York, 1977)

104. J. Herzog, A homological approach to symbolic powers, in *Commutative Algebra (Salvador, 1988)*. Lecture Notes in Mathematics, vol. 1430 (Springer, Berlin, 1990), pp. 32–44

105. J. Herzog, A generalization of the Taylor complex construction. Commun. Algebra **35**(5), 1747–1756 (2007)

106. J. Herzog, T. Hibi, *Monomial Ideals*. Graduate Texts in Mathematics, vol. 260 (Springer-Verlag London Ltd., London, 2011)

107. J. Herzog, A. Qureshi, Persistence and stability properties of powers of ideals. J. Pure Appl. Algebra **219**(3), 530–542 (2015)

108. J. Herzog, B. Ulrich, Self-linked curve singularities. Nagoya Math. J. **120**, 129–153 (1990)

109. J. Herzog, L.T. Hoa, N.V. Trung, Asymptotic linear bounds for the Castelnuovo-Mumford regularity. Trans. Am. Math. Soc. **354**(5), 1793–1809 (2002)

110. J. Herzog, T. Hibi, X. Zheng, Monomial ideals whose powers have a linear resolution. Math. Scand. **95**(1), 23–32 (2004)

111. J. Herzog, A. Rauf, M. Vladoiu, The stable set of associated prime ideals of a polymatroidal ideal. J. Algebraic Combin. **37**(2), 289–312 (2013)

112. H.T. Hien, H.M. Lam, N.V. Trung, Saturation and associated primes of powers of edge ideals. J. Algebra **439**, 225–244 (2015)

113. A. Hirschowitz, Une conjecture pour la cohomologie des diviseurs sur les surfaces rationelles génériques. J. Reine Angew. Math. **397**, 208–213 (1989)

114. L.T. Hoa, Stability of associated primes of monomial ideals. Vietnam J. Math. **34**(4), 473–487 (2006)

115. M. Hochster, C. Huneke, Comparison of symbolic and ordinary powers of ideals. Invent. Math. **147**(2), 349–369 (2002)

116. C. Huneke, Integral closures and primary components of ideals in three dimensional regular local rings. Math. Ann. **275**, 617–635 (1986)

117. A. Iarrobino, V. Kanev, *Power Sums, Gorenstein Algebras, and Determinantal Loci* (Springer, New York, 1999)

118. S. Jacques, Betti numbers of graph ideals. Ph.D. Thesis, University of Sheffield (2004). math.AC/0410107

119. I.B. Jafarloo, G. Zito, On the containment problem for fat points (2018). Preprint. arXiv:1802.10178

120. A.V. Jayanthan, S. Selvaraja, Asymptotic behavior of Castelnuovo-Mumford regularity of edge ideals of very well-covered graphs. J. Commut. Algebra. (to appear). arXiv:1708.06883
121. A.V. Jayanthan, S. Selvaraja, Upper bounds for the regularity of powers of edge ideals of graphs (2018). Preprint. arXiv:1805.01412
122. A.V. Jayanthan, N. Narayanan, S. Selvaraja, Regularity of powers of bipartite graphs. J. Algebraic Combin. **47**(1), 17–38 (2018)
123. T. Kaiser, M. Stehlík, R. Škrekovski, Replication in critical graphs and the persistence of monomial ideals. J. Combin. Theory Ser. A **123**, 239–251 (2014)
124. G. Kalai, R. Meshulam, Intersections of Leray complexes and regularity of monomial ideals. J. Combin. Theory Ser. A **113**(7), 1586–1592 (2006)
125. M. Katzman, Characteristic-independence of Betti numbers of graph ideals. J. Combin. Theory Ser. A **113**(3), 435–454 (2006)
126. K. Khashyarmanesh, M. Nasernejad, On the stable set of associated prime ideals of monomial ideals and square-free monomial ideals. Commun. Algebra **42**(9), 3753–3759 (2014)
127. V. Kodiyalam, Asymptotic behaviour of Castelnuovo-Mumford regularity. Proc. Am. Math. Soc. **128**, 407–411 (1999)
128. H.M. Lam, N.V. Trung, Associated primes of powers of edge ideals and ear decompositions of graphs. Trans. Am. Math. Soc. **372**(5), 3211–3236 (2019)
129. J.M. Landsberg, *Tensors: Geometry and Applications. Graduate Studies in Mathematics*, vol. 128 (American Mathematical Society, Providence, RI, 2012)
130. J.M. Landsberg, Z. Teitler, On the ranks and border ranks of symmetric tensors. Found. Comput. Math. **10**, 339–366 (2010)
131. R. Lazarsfeld, *Positivity in Algebraic Geometry I.–II*. Ergebnisse der Mathematik und ihrer Grenzgebiete, vols. 48–49 (Springer, Berlin, 2004)
132. L. Lovász, Normal hypergraphs and the perfect graph conjecture. Discrete Math. **2**(3), 253–267 (1972)
133. G. Malara, T. Szemberg, J. Szpond, On a conjecture of Demailly and new bounds on Waldschmidt constants in $\mathbb{P}^N$. J. Number Theory **189**, 211–219 (2018)
134. P. Mantero, The structure and free resolution of the symbolic powers of star configurations of hypersurfaces (2019). Preprint. arXiv:1907.08172
135. J. Martinez-Bernal, S. Morey, R. Villarreal, Associated primes of powers of edge ideals. Collect. Math. **63**(3), 361–374 (2012)
136. S. McAdam, *Asymptotic Prime Divisors*. Lecture Notes Mathematics, vol. 1023 (Springer, Berlin, 1983)
137. E. Miller, B. Sturmfels, *Combinatorial Commutative Algebra*. Graduate Texts in Mathematics, vol. 227 (Springer, New York, 2004)
138. N.C. Minh, N.V. Trung, Cohen-Macaulayness of monomial ideals and symbolic powers of Stanley-Reisner ideals. Adv. Math. **226**, 1285–1306 (2011)
139. M. Moghimian, S.A. Seyed Fakhari, S. Yassemi, Regularity of powers of edge ideal of whiskered cycles. Commun. Algebra **45**(3), 1246–1259 (2017)
140. S. Morey, Depths of powers of the edge ideal of a tree. Commun. Algebra **38**(11), 4042–4055 (2010)
141. M. Nagata, On the 14-th problem of Hilbert. Am. J. Math. **81**, 766–772 (1959)
142. M. Nasernejad, K. Khashyarmanesh, On the Alexander dual of the path ideals of rooted and unrooted trees. Commun. Algebra **45**(5), 1853–1864 (2017)
143. E. Nevo, I. Peeva, C_4-free edge ideals. J. Algebraic Combin. **37**(2), 243–248 (2013)
144. P. Norouzi, S.A. Seyed Fakhari, S. Yassemi, Regularity of powers of edge ideals of very well-covered graphs (2017). Preprint. arXiv:1707.04874
145. J.P. Park, Y.S. Shin, The minimal free resolution of a star-configuration in $\mathbb{P}^n$. J. Pure Appl. Algebra **219**, 2124–2133 (2015)
146. I. Peeva, *Graded Syzygies*. Algebra and Applications, vol. 14 (Springer-Verlag London Ltd., London, 2011)
147. K. Ranestad, F. Schreyer, Varieties of sums of powers. J. Reine Angew. Math. **525**, 147–181 (2000)

148. L.J. Ratliff Jr., A brief survey and history of asymptotic prime divisors. Rocky Mountain J. Math. **13**, 437–459 (1983)

149. E. Scheinerman, D. Ullman, *Fractional Graph Theory. A Rational Approach to the Theory of Graphs* (Wiley, New York, 1997)

150. B. Segre, Alcune questioni su insiemi finiti di punti in geometria algebrica. Atti Convegno Intern. di Geom. Alg. di Torino (1961), pp. 15–33

151. S.A. Seyed Fakhari, S. Yassemi, Improved bounds for the regularity of powers of edge ideals of graphs (2018). Preprint. arXiv:1805.12508

152. P. Shenzel, Examples of Gorenstein domains and symbolic powers of monomial space curves, J. Pure Appl. Algebra **71**, 297–311 (1991)

153. A. Simis, W. Vasconcelos, R.H. Villarreal, On the ideal theory of graphs. J. Algebra **167**(2), 389–416 (1994)

154. H. Skoda, Estimations L^2 pour l'opérateur $\widehat{\partial}$ et applications arithmétiques, in *Séminaire P. Lelong (Analyse), 1975/76*. Lecture Notes Mathematics, vol. 578 (Springer, New York, 1977), pp. 314–323

155. R. Stanley, *Combinatorics and Commutative Algebra*. Progress in Mathematics, vol. 41 (Birkhäuser Boston, Inc., Boston, MA, 1983)

156. B. Sturmfels, S. Sullivant, Combinatorial secant varieties. Pure Appl. Math. Q. **2**(3), Part 1, 867–891 (2006)

157. S. Sullivant, Combinatorial symbolic powers. J. Algebra **319**(1), 115–142 (2008)

158. I. Swanson, Linear equivalence of topologies. Math. Z. **234**, 755–775 (2000)

159. I. Swanson, Primary decompositions, in *Commutative Algebra and Combinatorics*. Ramanujan Mathematical Society Lecture Notes Series, vol. 4 (Ramanujan Mathematical Society, Mysore, 2007), pp. 117–155

160. Y. Takayama, Combinatorial characterizations of generalized Cohen-Macaulay monomial ideals. Bull. Math. Soc. Sci. Math. Roumanie (N.S.) **48**, 327–344 (2005)

161. N. Terai, N.V. Trung, On the associated primes and the depth of the second power of squarefree monomial ideals. J. Pure Appl. Algebra **218**(6), 1117–1129 (2014)

162. N.V. Trung, H. Wang, On the asymptotic behavior of Castelnuovo-Mumford regularity. J. Pure Appl. Algebra **201**(1–3), 42–48 (2005)

163. J. Vallès, Fibrés logarithmiques sur le plan projectif. Ann. Fac. Sci. Toulouse Math. **16**(2), 385–395 (2007)

164. A. Van Tuyl, A beginner's guide to edge and cover ideals, in *Monomial Ideals, Computations and Applications*. Lecture Notes Mathematics, vol. 2083 (Springer, New York, 2013), pp. 63–94

165. W. Vasconcelos, Symmetric algebras and factoriality, in *Commutative Algebra, Berkeley, CA, 1987)*. MSRI Publications, vol. 15 (Springer, New York, 1989), pp. 467–496

166. R.H. Villarreal, *Monomial Algebras*. Monographs and Textbooks in Pure and Applied Mathematics, 2nd edn. (CRC Press, New York, 2015)

167. M. Waldschmidt, Propriétés arithmétiques de fonctions de plusieurs variables. II, in *Séminaire P. Lelong (Analyse), 1975/76*. Lecture Notes Mathematics, vol. 578 (Springer, New York, 1977), pp. 108–135

168. R.M. Walker, Uniform Harbourne-Huneke Bounds via Flat Extensions. J. Algebra **516**, 125–148 (2018)

169. G. Wegner, d-collapsing and nerves of families of convex sets. Arch. Math. (Basel) **26**, 317–321 (1975)

170. G. Whieldon, Stabilization of Betti tables. J. Commut. Algebra **6**, (1), 113–126 (2014)

171. R. Woodroofe, Matchings, coverings, and Castelnuovo-Mumford regularity. J. Commut. Algebra **6**(2), 287–304 (2014)

172. O. Zariski, P. Samuel, *Commutative Algebra*, vol. II (Springer, New York, 1960)

173. X. Zheng, Resolutions of facet ideals. Commun. Algebra **32**, 2301–2324 (2004)

Index

Symbols
$\alpha(I)$, 73

A
Annihilator, 4, 14, 28, 121
Apolarity, 121, 123
    action, 121
    Apolarity Lemma, 123, 125
Associated graded ring, 8
Associated primes, 28, 29, 31
    ideal, 3, 5
    module, 5
    set of, 5

C
Chromatic number, 20, 26
    b-fold, 87
    fractional, 87, 88
Clique, 20
Co-chordal number, 53
Cohen-Macaulay, 42
Collaboration, 147
Colouring, 20
    b-fold, 87
Complete intersection, 80
Containment problem, 71, 77, 97
Cover ideal, 18, 21, 25, 26, 28, 29, 31, 139, 140
Critically chromatic, 27, 28, 31

D
Deficiency, 22–24
Degree complex, 48
Deletion, 41
Duplication, 22

E
Edge ideal, 18, 19, 21, 24, 44, 45, 51–55, 57–60, 64, 86, 88, 138, 139
Even-connected, 58
Exceptional curve, 104
Experiments, 144

F
Fat point, 72, 141, 142
    degree, 72
Fekete's Subadditivity Lemma, 74
Forbidden locus, 129, 130

G
Generic element, 117
Generic Hilbert function, 97
Graph, 18, 43
    bipartite, 20, 21, 53
    chordal, 43, 52, 140
    clique, 20, 55
    co-chordal, 54
    complement, 19, 52

LECTURE NOTES OF THE UNIONE
MATEMATICA ITALIANA

 **Springer**

Editor in Chief: Ciro Ciliberto and Susanna Terracini

Editorial Policy

1. The UMI Lecture Notes aim to report new developments in all areas of mathematics and their applications - quickly, informally and at a high level. Mathematical texts analysing new developments in modelling and numerical simulation are also welcome.

2. Manuscripts should be submitted to
   Redazione Lecture Notes U.M.I.
   umi@dm.unibo.it
   and possibly to one of the editors of the Board informing, in this case, the Redazione about the submission. In general, manuscripts will be sent out to external referees for evaluation. If a decision cannot yet be reached on the basis of the first 2 reports, further referees may be contacted. The author will be informed of this. A final decision to publish can be made only on the basis of the complete manuscript, however a refereeing process leading to a preliminary decision can be based on a prefinal or incomplete manuscript. The strict minimum amount of material that will be considered should include a detailed outline describing the planned contents of each chapter, a bibliography and several sample chapters.

3. Manuscripts should in general be submitted in English. Final manuscripts should contain at least 100 pages of mathematical text and should always include

   – a table of contents;
   – an informative introduction, with adequate motivation and perhaps some historical remarks: it should be accessible to a
     reader not intimately familiar with the topic treated;
   – a subject index: as a rule this is genuinely helpful for the reader.

4. For evaluation purposes, please submit manuscripts in electronic form, preferably as pdf- or zipped ps-files. Authors are asked, if their manuscript is accepted for publication, to use the LaTeX2e style files available from Springer's web-server at
   ftp://ftp.springer.de/pub/tex/latex/svmonot1/ for monographs
   and at
   ftp://ftp.springer.de/pub/tex/latex/svmultt1/ for multi-authored volumes

5. Authors receive a total of 50 free copies of their volume, but no royalties. They are entitled to a discount of 33.3% on the price of Springer books purchased for their personal use, if ordering directly from Springer.

6. Commitment to publish is made by letter of intent rather than by signing a formal contract. Springer-Verlag secures the copyright for each volume. Authors are free to reuse material contained in their LNM volumes in later publications: A brief written (or e-mail) request for formal permission is sufficient.

Printed in the United States
By Bookmasters